Jeff Bezos

Best Quotes of the Richest Man on Earth

(Biography of a Billionaire Business Titan Jeff Bezos and the Everything Store)

Willard Jepson

Published By **Phil Dawson**

Willard Jepson

All Rights Reserved

Jeff Bezos: Best Quotes of the Richest Man on Earth (Biography of a Billionaire Business Titan Jeff Bezos and the Everything Store)

ISBN 978-1-7752884-8-0

No part of this guidebook shall be reproduced in any form without permission in writing from the publisher except in the case of brief quotations embodied in critical articles or reviews.

Legal & Disclaimer

The information contained in this book is not designed to replace or take the place of any form of medicine or professional medical advice. The information in this book has been provided for educational & entertainment purposes only.

The information contained in this book has been compiled from sources deemed reliable, and it is accurate to the best of the Author's knowledge; however, the Author cannot guarantee its accuracy and validity and cannot be held liable for any errors or omissions. Changes are periodically made to this book. You must consult your doctor or get professional medical advice before using any of the suggested remedies, techniques, or information in this book.

Upon using the information contained in this book, you agree to hold harmless the Author from and against any damages, costs, and expenses, including any legal fees potentially resulting from the application of any of the information provided by this guide. This disclaimer applies to any damages or injury caused by the use and application, whether directly or indirectly, of any advice or information presented, whether for breach of contract, tort, negligence, personal injury, criminal intent, or under any other cause of action.

You agree to accept all risks of using the information presented inside this book. You need to consult a professional medical practitioner in order to ensure you are both able and healthy enough to participate in this program.

Table Of Contents

Chapter 1: Family Background And Childhood 1

Chapter 2: A Visionary Mindset 7

Chapter 3: The Birth Of Amazon 15

Chapter 4: Scaling The Heights............... 23

Chapter 5: The Dot-Com Bubble And Beyond... 33

Chapter 6: A Company Transformed 42

Chapter 7: The Disruptor's Mindset 50

Chapter 8: Customer Obsession And Innovation Building A Culture Of Customer-Centricity................................ 59

Chapter 9: Chasing Moon Shots 66

Chapter 10: Challenges And Criticisms... 74

Chapter 11: Personal Life And Philanthropy 83

Chapter 12: The Washington Post And More 89

Chapter 13: Wealth, Influence, And Legacy 97

Chapter 14: Transition And New Beginnings 104

Chapter 15: Closing Thoughts 112

Chapter 16: Modest Beginnings 116

Chapter 17: Emergence Of An Empire.. 124

Chapter 18: Stingy Philanthropist 132

Chapter 19: What Lies Ahead............... 141

Chapter 20: Who Is Jeff Bezos? 163

Chapter 21: Business Career 171

Chapter 1: Family Background And Childhood

Jeffrey Preston Bezos, widely called Jeff Bezos, was born on the 12th of January 1964 at Albuquerque, New Mexico, USA. Bezos is an acclaimed as a media entrepreneur, an entrepreneur and commercial spaceman. Bezos is well-known as the entrepreneur behind Amazon the online retailer which grew into one of the largest and most influential companies in technology.

Family Background:

Jeff Bezos was born to the diverse family of. The mother of Jeff, Jacklyn Gise Jorgensen, was just a teenager in the year of his birth and was married to his father of birth, Ted Jorgensen, when she was 17. Their marriage, however, was not long-lasting however, and Jacklyn was remarried to Miguel Bezos, a Cuban immigrants, at the time that Jeff was just four years old. young. Miguel took in Jeff who he adopted, taking his name "Bezos," which he has been known as to this day.

Childhood and Early Years:

Bezos had an early fascination with the field of engineering and technology. He spent a lot of his early years playing around with machines and gadgets. He also transformed the garage that his parents had built into a lab where it was where he conducted various scientific experiments. Bezos his interest in space exploration and exploration developed during his early years.

Bezos was a student at Miami Palmetto High School in Florida in the United States, where he performed exceptionally academically. He also demonstrated an early ability to be an entrepreneur. He was a participant in the Student Science Training Program at the University of Florida, where was conducting research on the subject that dealt with artificial intelligence.

Following his high school graduation, Bezos enrolled at Princeton University and was awarded a summa cum loupe in 1986. He earned a diploma in electrical engineering and computer science. While at Princeton as a student, he was part of the Phi Beta Kappa society and the Princeton Chapter of Students for the Exploration and Development of Space.

Career Beginnings:

Following his graduation in Princeton, Bezos worked in different fields of technology. He was employed by several businesses, including Fitel, Bankers Trust, as well as D.E. Shaw & Co., which was a hedge fund. It was while he worked working at D.E. Shaw when he was captivated by the possibilities of the internet's emergence.

It was 1994 when Bezos took the plunge to walk away from his lucrative job at D.E. Shaw in order to pursue his business plans. He relocated to Seattle and established Amazon.com, a bookstore online which aimed at leveraging the rising popularity of web for shopping.

Conclusion:

Jeff Bezos' family background and early childhood had a significant impact in forming his initial interests and predisposition to technological advancement, science and entrepreneurialism. The early exposure to a variety of experiences, along with the solid foundation of his education and interest in the web, eventually led to the development of Amazon the company which changed the way that people buy and established the basis for his huge growth in the technology industry.

Educational Journey and Early Interests

Jeff Bezos' educational journey and his early pursuits were a major factor in developing his business mindset and business approach. Below is a fuller review:

1. High School and Early Curiosity:

Bezos has attended Miami Palmetto High School in Florida. Through the time he was in high school Bezos showed a keen fascination with technology and innovating. Participated in engineering and science fairs showing his interest and ability to think on his feet.

2. Princeton University:

Following high school, Bezos took a course at Princeton University, where he completed a bachelor's master's degree in electrical engineering as well as computer science. The time he spent at Princeton provided him with a challenging academic setting as well as provided him with an excellent foundation in the fields of science and technology. While at Princeton He continued to demonstrate his enthusiasm for research and research.

3. Early Career and Transition to Finance:

After his college graduation at Princeton, Bezos worked in the tech industry. He held posts at firms such as Fitel as well as Bankers Trust. But he switched to finance and joined

D.E. Shaw & Co., an eminent hedge fund. Even after his move to finance, his specialized experience and passions continue to shape his outlook regarding commercial opportunities.

4. Internet's Potential:

While at D.E. Shaw, Bezos became increasingly fascinated by the explosive expansion of the internet and the potential for it to transform many industries, especially commerce. Bezos realized that the web could be used as a tool for new business models and strategies that are geared towards customers.

Chapter 2: A Visionary Mindset

The Formation of Bezos' Philosophical Principles

Jeff Bezos' philosophical principles are a major factor in determining his attitude to business, innovation, as well as his leadership. The principles have proved to be a key factor in the direction of his choices and driving the growth of Amazon, as well as influencing his outlook regarding various areas of his life. A few of the key elements in the evolution of Bezos philosophy include:

1. Early Influences and Experiences:

Bezos His upbringing, his training, and his exposure to different experience have all led to the development of his philosophy. He was attracted to technology from a young age and his studies in Princeton University, and his encounters in diverse sectors like finance and technology each played an important role in developing his philosophy.

2. Long-Term Vision:

One of Bezos his fundamental values is his commitment to longer-term thinking. Bezos has always advocated focus on the long-term and avoiding the demands of immediate gain. This outlook is founded on his belief that real change and impactful innovation requires the willingness to invest in and risk their lives in the interest of further growth and development.

3. Customer Obsession:

Bezos his customer-focused approach is the foundation of his thought process. The company's performance is directly linked to its capacity to satisfy the needs of customers and surpass them. This belief has led to Amazon's dedication to providing excellent customer experience and constantly expanding its offerings and services.

4. Innovation and Experimentation:

Bezos is a proponent of a culture that encourages ingenuity and experimentation at Amazon. He believes that failing is aspect of success, and that a willingness to take sensible risks is crucial to stimulating the pace of innovation. This is the reason why this principle has allowed Amazon to test new concepts and to enter markets far beyond the original market of an online retailer.

5. Embracing Change:

Bezos is known as a man who embraces changes and disruptions instead of avoiding the possibility of. The company's business evolves as technology advances, while the preferences of customers change. Instead of being resistant to the change, he helps his employees to stay in the forefront and be able to adapt to the changing environment.

6. Relentless Pursuit of Excellence:

Bezos's pursuit of excellence is apparent in his very high standards of Amazon's products, operations, and services. He is a firm believer in setting lofty objectives and continually seeking to achieve the goals. This philosophy has fueled Amazon's innovations in fields like cloud computing, logistics (Amazon Web Services) as well as other areas.

7. Diverse Thinking and Decision-Making:

Bezos believes in diversity of viewpoints and promotes dialogue in Amazon. He believes that a broad perspectives could lead to more informed decisions and better solutions. This is the reason for Amazon's broad range of products and services.

8. Willingness to Learn and Adapt:

Despite his huge achievement, Bezos maintains a willingness to grow and learn. He is always seeking out new knowledge and listens to critique and adjusts his strategy as needed. The humility and willingness towards growth is a key part of his philosophy.

As a summary, Jeff Bezos' philosophical ideas have been developed by the combination of his own experience, knowledge, as well as the difficulties he encountered in his role as an businessman. A focus on the long-term and customer-centricity, as well as innovation and adaptability have been a major influence on the direction of Amazon but also played an impact in the way businesses strategy for growth and development.

Early Career and Transition to Wall Street

Jeff Bezos' early career and the transition into Wall Street played a crucial influence on his entrepreneurial skills and laying the foundation for his entrepreneur journey. The following is an in-depth look at the time period:

1. Education and Early Interests:

Following his graduation from Princeton University in 1986 with an electrical engineering degree and computer science Bezos considered a variety of career options which allowed him to use his expertise in technology and drive for ingenuity.

2. Fitel and Bankers Trust:

Bezos his first position after university was working at Fitel the company with the goal of creating an international network of trade. The company was working on a system of computers to manage trade and settlements.

In the following years, Bezos moved to Bankers Trust which is a finance firm. He worked there on the development of software that could automate trading as well as other financial transactions. Although his work was mainly tech-related, the exposure to the finance industry as well as his contacts in the company of Wall Street professionals would later be beneficial.

3. D.E. Shaw & Co.:

In the year 1990, Bezos made a significant change in his career by joining D.E. Shaw & Co., an investment firm that is recognized for its innovative algorithmic use in investing and trading. Bezos was initially an executive vice president, however he quickly climbed up the ranks because of his technological skills as well as his strategic planning.

At D.E. Shaw, Bezos led a group of developers who developed algorithms that could identify and capitalize on the market's developments. The experience did not just enhance his knowledge of finance, but it also introduced Bezos to the possibilities of analysis and automation concepts, which would eventually be central to the Amazon's business.

4. Recognition of E-Commerce Potential:

In his time at D.E. Shaw, Bezos began to recognize the immense potential of the internet to commerce. Bezos saw the way that the internet can disrupt the traditional model of retail and provide new possibilities to grow and innovate.

5. The Birth of Amazon:

Bezos his realization of the possibilities of the internet prompted him to take a risky venture. In 1994, Bezos quit the high-paying position at D.E. Shaw and created Amazon.com which is an online store. Shaw chose to use books as a base because they were merchandise that had a wide appeal and could easily be sold and delivered online.

Bezos His experiences in Wall Street undoubtedly influenced his business strategy. Experience with analytical analysis and data-driven decision-making as well as risk management at D.E. Shaw helped him to take calculated business decisions and control Amazon's rapid growth with a strategic approach.

In summation, Jeff Bezos' early career, with his stints in Fitel, Bankers Trust, and most notably D.E. Shaw & Co., gave him a distinctive blend of technical knowledge as well as financial insight and a deep understanding of methods based on data. This experience gave him the basis he needed to understand and harness the potential for transformation of the web, which led to the birth of Amazon and the subsequent recognition in the ranks of top entrepreneurs worldwide.

Chapter 3: The Birth Of Amazon

The Concept that ignited It All

The beginnings of Amazon can be traced to a basic but radical concept that Jeff Bezos had. It was this idea that sparked the development of the online retailer that we have now. The concept was to capitalize on the growing internet and capitalize on its power to change how people buy and sell their products. This is the tale of how the concept became a reality:

1. The Insightful Realization:

In the beginning of 1990, Jeff Bezos was working at D.E. Shaw & Co., a quantitative hedge fund. It was during his tenure that Shaw & Co. witnessed the explosive expansion of the internet as well as its potential to change conventional industries. He began to be particularly fascinated by the idea that the web could become a medium for commerce that could provide a different method for customers to buy items and also for companies to connect with clients.

2. The 1994 Cross-Country Drive:

Bezos his idea began to form as he drove across the country beginning in New York City to Seattle. He wanted to make a shift and was interested in exploring possibilities for entrepreneurship. On the way He came across a number of possible business ideas, however the concept of launching an online bookshop caught his attention.

3. Why Books?

Bezos realized that books are an item with global popularity, an array of possibilities and an easy model for selling online. He realized that selling books online could overcome the restrictions of traditional retail shops, which were limited by space on shelves and the cost of inventory. Furthermore, he realised that the market for books was huge enough for a dedicated online marketplace.

4. The 3 Key Factors:

Bezos looked at three elements that helped make books the ideal platform for his venture online:

The books are available with a wide range of titles, genres and languages, providing an array of options for readers.

Global Reach: The books can be enjoyed by all and are therefore suitable to be read by a wide public.

Logistical Feasibility: As opposed to heavy or fragile objects the books were relatively simple to pack, store and then ship.

5. The Birth of Amazon:

With this idea in the back of his mind, Bezos decided to pursue his idea. The company relocated to Seattle for the convenience of being near the technology hub, and also to benefit from the lack of sales taxes, which is advantageous to the online retail industry. In the summer of 1994, he launched Amazon.com to be an online retailer.

6. The Evolution and Beyond:

Although Amazon began as a bookshop, Bezos' visionary approach made him see the company as a platform which could be used to sell virtually anything on the internet. Through the years, Amazon expanded its offerings to cover a range of segments, launched services such as Amazon Prime, and even entered into other industries such as cloud computing (Amazon Web Services).

The genesis of Amazon was due to Jeff Bezos' keen observation of the possibilities of the internet, and his ingenious idea of using it to facilitate online shopping. The shrewdness of his decision to select books for the first item segment, along with his emphasis on increasing customer satisfaction and ensuring long-term expansion, set the stage for the transformation of Amazon into one of the biggest and most powerful companies.

The Garage Days: Amazon's Humble Beginnings

Amazon's humble beginnings are usually linked to it's "garage days," a time when Amazon was only starting out and was operating in a garage located in Bellevue, Washington. The early years set the conditions for Amazon's amazing transformation from an bookshop to a worldwide technological and e-commerce powerhouse.

1. The Garage Setup:

When Jeff Bezos officially founded Amazon.com in July 1994, he first managed the company from his home located in Bellevue. It was an office in a temporary setting, and Bezos together with his team created the foundation of the store's online.

2. DIY Spirit:

In this period, the environment was typical like a typical start-up's bustle and ingenuity. They were an DIY approach to everything, from packing orders all the way to writing codes. Bezos himself would put books into cardboard boxes and take these boxes over to the Post Office in order to ship them.

3. The Desk as Symbolism:

A famous image that dates back to the beginning of Amazon's existence includes a photo that shows Jeff Bezos sitting at his desk. It was a door made of wood supported by sawhorses. The desk, often called"the "door desk," became an iconic symbol of Amazon's flimsy start and its commitment to efficiency.

4. Focus on Books:

At its beginning, Amazon focused solely on selling books on the internet. Bezos and his team was meticulous in entering details about the books they sold, developed page pages for products, and worked to enhance the experience for customers. They saw the potential of the web to provide an enormous selection of books, which physical bookstores were unable to match.

5. Expansion and Adaptation:

When Amazon was gaining traction and customers confidence, it started expanding its product offerings to include more than the books. Its ability to rapidly change and invent led it to discover new markets as well as design its own product lines including the Kindle electronic reader.

6. Challenges and Perseverance:

The early days of the garage were not without obstacles. Amazon was faced with financial pressures as well as competition from traditional retail stores, and technological obstacles. But, Bezos and his team was determined to conquer the obstacles they faced and remain true to their long-term goals.

7. Legacy of the Garage Days:

The days of the garage at Amazon are a symbol of its beginnings as well as its entrepreneurial spirit. The idea of starting with an unassuming space before expanding to become a global company is a common theme for entrepreneurs who are aspiring. Amazon's rise from a garage to becoming a world-class technology giant demonstrates the strength of creativity, focus on customers and perseverance.

It is clear that Amazon's humble start in a garage reflects the commitment, dedication and a creative mindset that set the stage of the company's phenomenal expansion. Jeff Bezos' willingness to begin with a small scale and then be prepared for risky decisions, along with the company's commitment to delivering a great experience for customers, set the course for the company's transformation to become one of the most powerful companies worldwide.

Chapter 4: Scaling The Heights

Navigating Initial Challenges and Setbacks

Amazon's path towards "scale the heights" and become an international e-commerce and tech major wasn't free of setbacks and challenges. Despite overcoming numerous challenges and challenges, Amazon's ability to meet these challenges and adjust played an important part in the success it achieved. Let's take a look at a few of the first challenges and back-slashes Amazon faced during its development:

1. Financial Strain:

At the beginning of its existence, Amazon faced financial pressure when it poured a lot of money into construction of infrastructure, technology and stock. It was burning up money and had concerns regarding its capacity to make into a profit. Bezos was known for his emphasis on long-term growth rather than short-term profit and this strategy was successful, but it also resulted in skepticism by investors.

2. Intense Competition:

Amazon was a newcomer to the market, competing from brick-and-mortar stores that are well-established and others online retailers. In order to compete with traditional retailers, Amazon to stand out by offering ease of use, variety, and price that was competitive.

3. Technological Hurdles:

Maintaining and enhancing a strong infrastructure for technology was essential to Amazon's business. Amazon was required to continuously develop its website, enhance customer experience and resolve issues with technology to provide the smooth shopping experience of its customers.

4. Inventory Management:

The efficient management of inventory was the challenge of managing inventory efficiently, especially when Amazon increased its range of products. The balance between demand and supply while avoiding overstocking or understocking and maximizing distribution centers are essential for efficient operations.

5. Customer Trust and Satisfaction:

The beginning of the internet was met with distrust from consumers that were wary of the sharing of personal information and financial details on the web. Amazon was required to earn confidence through safe transactions, dependable delivery as well as a superior customer support.

6. Operational Scalability:

With the rise of Amazon's popularity, the company needed to increase operations quickly to keep up with the demand. This meant expanding its distribution channels and improving its fulfillment process as well as ensuring that deliveries are timely and ensuring quality.

7. Diversification and Adaptation:

Amazon's expansion into different product categories that went beyond books demanded thorough market research as well as adapting to the specific demands of each product type of product. The diversification of the product line required understanding various types of customers, and tailoring the user experience to suit each one.

8. Dot-com Bubble and Resilience:

In the latter part of 1990, the dot-com bubble broke, causing the demise of many companies that were based on the internet. Amazon was viewed with suspicion regarding its future viability, and was required demonstrate its strength. Bezos determination to stick with a the long term, in conjunction with cuts in costs allowed Amazon to stand up against this challenging time.

While navigating through these difficulties, Jeff Bezos and his team displayed a mix of creative thinking, strategic planning as well as the determination to take prudent risks and a commitment to ensuring customer satisfaction. Instead of getting discouraged from setbacks Amazon made use of these challenges as an opportunity to improve, learn to improve the efficiency of its operation.

The ability of Amazon to get over initial hurdles and setbacks is a testimony to its determination, customer-centered strategy, and dedication to continuous improvement. Amazon's rise from a tiny online store into a world-class e-commerce company and technology company demonstrates the resilience of its employees and perseverance even in the face of hardship.

Initial Growth Strategies as well as Expansion Plans of Amazon's initial growth strategies and expansion played a key role in propelling the business from being a small, online bookshop into a worldwide e-commerce and technological giant. Jeff Bezos and his team utilized innovative strategies that created the conditions for Amazon's explosive growth as well as its transformation into various service and product lines. Here are some of the key strategies and plans for expansion:

1. Customer-Centric Approach:

At the beginning of its existence, Amazon placed a strong focus on customer satisfaction. Amazon's emphasis on delivering an effortless shopping experience, efficient delivery and outstanding customer service has helped to build the trust and confidence of the customers it serves.

2. Diverse Product Expansion:

Following its success as an e-commerce company that is a success, Amazon strategically expanded into various products. The expansion was fueled by the use of data to make decisions along with market research, as well as discovering customer demands. The firm's "Everything Store" vision led to the addition of an array of items that range from electronic gadgets to clothing and household items.

3. Third-Party Sellers and Marketplace:

In the year 2000, Amazon launched its third-party seller platform that allowed the independent seller to advertise and sell their products through Amazon's platform. This strategy for marketplaces significantly expanded the range of goods that customers could purchase, while helping to increase the revenue of Amazon by way of commissions and charges.

4. Amazon Prime:

The service was launched in 2005. Amazon Prime revolutionized the e-commerce market. The subscription program offered customers quick and fast shipping on items that were eligible and also access to streaming media, among additional advantages. Amazon Prime incentivized customer loyalty and encouraged customers to make repeat purchases.

5. Innovation and Technology:

Amazon has always put money into innovative technology and innovations to boost its efficiency. It has also improved the algorithms that recommend products, improving the user experience on its site as well as developing its cloud computing services, Amazon Web Services (AWS).

6. Fulfillment Network:

Amazon established a large system of distribution centers as well as fulfillment centers in order to guarantee the efficient storage of inventory, order processing and on time delivery. This helped Amazon cut down on shipping time and increase its footprint globally.

7. International Expansion:

Once it established itself as a major player within Amazon's home country of the United States, Amazon began expanding its reach internationally. Amazon gradually introduced countries-specific sites, localizing the products to meet diverse markets and different cultural backgrounds.

8. Acquisitions and Ventures:

Amazon was pursuing strategic acquisitions and initiatives to diversify its offering and broaden its reach. Some notable acquisitions include Zappos (an online retailer of shoes and apparel retailer) as well as Twitch (a live streaming service specifically for gamers).

9. Kindle and E-Books:

Amazon changed the way publishers publish through the launch of the Kindle ebook reader in 2007. This revolutionary device enabled users to read and download books electronically, which changed how people consume printed media.

10. AWS and Cloud Computing:

Amazon's move into cloud computing through Amazon Web Services (AWS) was an important growth catalyst. AWS offered businesses cloud-based infrastructure and service, which became the main source of revenue for Amazon.

The gist of Amazon's initial expansion strategies were centered around the importance of customer service, diversification of products as well as technological advancement as well as global expansion. Jeff Bezos' willingness to make calculated risk, along with Amazon's constant determination to be the best, set the groundwork for Amazon's remarkable transformation from a modest garage to becoming one of the most influential and successful businesses.

Chapter 5: The Dot-Com Bubble And Beyond

Surviving the Dot-Com Crash

Amazon's capability to survive and flourish after the dot-com market crisis is proof of its ability to adapt, resilience and strategic decisions. The dot-com bubble that was bursting in the first quarter of 2000 was a time when many companies on the internet fall, but Amazon was able to survive the storm, and emerge an even stronger and concentrated firm. This is the way Amazon was able to navigate the maze of the dot-com bust:

1. Focus on Long-Term Strategy:

In the midst of the dot-com bubble Jeff Bezos and his team continued to maintain a long-term outlook. Amazon's primary mission and its determination to provide customers with the best service was in a constant state. Instead of pursuing short-term profits Amazon invested in innovation in technology and expand its product offerings.

2. Cost-Cutting Measures:

As a response to the downfall of dot-com, Amazon implemented cost-cutting measures to save resources. Amazon tightened its belts through reducing costs, enhancing the operations and revising the business practices it follows.

3. Diversification of Product Categories:

When Amazon was initially a web-based store, it has expanded into other products that went beyond books. This diversification aided in reducing the impact of the market crash. The wide variety of products offered by Amazon helped it become less dependent of a single line of products and helped the company gain a wider client base.

4. Amazon Web Services (AWS):

The introduction by Amazon to Amazon Web Services (AWS) at the end of 2006 marked a paradigm shift. AWS provided cloud computing services to companies, which allowed companies to outsource their IT infrastructures to Amazon's servers. This not only brought in substantial revenue, but also highlighted the technological innovations and flexibility of Amazon.

5. Financial Prudence:

Amazon's prudent financial management in the dot-com slump helped keep its the stability. Amazon's primary focus was on producing positive cash flow while avoiding excess debt helped it stand out in comparison to its more cash-strapped rivals.

6. Strong Brand and Customer Trust:

Amazon's focus on providing top customer service as well as reliable delivery created a solid brand image and gained the confidence of its clients. This trust was an important resource during times of uncertainty which helped Amazon retain its loyal customers even when other platforms for e-commerce have failed.

7. Reduced Reliance on Venture Capital:

In contrast to many of the dot-com companies that heavily depended on venture capital, Amazon had pursued a prudent method of finance. With its gradual growth rate and self-sustaining business model meant it wasn't as susceptible to the fluctuations of market conditions.

8. Innovation and Adaptation:

In its entire time, Amazon had demonstrated a capacity to be innovative and adapt to changes in the market. This ability was essential to surviving the dot-com bust. The company continuously improved its processes, technology as well as its operations to accommodate changing customer demands.

After the dot-com boom, Amazon's strategy-driven decisions as well as its capability to implement its long-term strategy resulted in sustained expansion and the success. Its focus on the customer's needs, diversification and stability in its finances has not only helped it withstand the financial crisis, but also helped position Amazon as an industry leader in the technology and e-commerce sectors for the foreseeable future.

Diversification and Innovation Amidst Turmoil

Innovation and diversification was a key factor in Amazon's ability in uncertain times and emerge more durable and stronger firm. Amazon's dedication to exploring new possibilities, opening up to different markets and taking on the latest technologies helped it conquer obstacles and continue its upward trend. This is the ways in which Amazon was able to embrace diversification and change amid turmoil

1. Diversification of Product Categories:

The initial successes of Amazon as an online retailer served as a basis for the expansion of its business into other segments of products. Amazon's offerings were diversified to include electronic products, apparel household goods, other categories. Its diversification decreased its dependence on a single line of products and made it more resilient to shifts in market.

2. Amazon Web Services (AWS):

One of the most ingenuous Amazon actions was the launch the concept of Amazon Web Services (AWS). Being the first to pioneer cloud computing AWS offered businesses the ability to access computing resources on demand and also services. The expansion beyond e-commerce was not solely brought substantial revenues, it additionally demonstrated Amazon's technical expertise and flexibility.

3. Third-Party Marketplace:

Amazon's third party marketplace has permitted independent sellers to sell their items on the marketplace. This change not only increased the selection of goods that customers could purchase, but it additionally created a brand new source of revenue for Amazon by way of commissions and charges.

4. Acquisitions and Ventures:

Amazon's acquisitions such as Zappos (online retail store for shoes) as well as Twitch (live streaming site) have demonstrated the company's willingness to explore new markets and pursue new ventures. These acquisitions widened Amazon's reach and diversified its offerings.

5. Innovation in Fulfillment and Logistics:

Amazon changed the way logistics is handled through the introduction of technologies such as robotic warehouses, modern system for picking orders, and next-day delivery. These advancements improved effectiveness, decreased costs, as well as improved customer service.

6. Kindle and E-Books:

The release of the Kindle electronic reader was an extraordinary invention that led to a revolution in the publishing business. The move by Amazon into electronic books allowed it to manage the consumption and distribution of digital content. It also opened opportunities for new sources of revenue.

7. Amazon Prime and Subscription Models:

The launch of Amazon Prime, with its speedy and no-cost shipping, as well as extra benefits, led to customers to loyalty and make frequent purchases. Amazon's move into subscription models to various services also broadened its sources of revenue.

8. Focus on Customer-Centric Innovation:

Amazon's unwavering focus on satisfying customers' needs with innovation sets the company distinct. Amazon relied on feedback from customers as well as data analytics and technology to continually improve its shopping experience from personal suggestions to easy return policies.

9. Investment in Original Content:

Amazon's entrance into the world of entertainment with Amazon Prime Video showcased its determination to innovate.

Amazon invested heavily in the production of television and original films expanding its range of offerings as well as expanding its range of offerings.

Overall, Amazon's diversification as well as its commitment to innovating has allowed it to endure the turbulence, but also prosper under challenging conditions. In exploring new markets and making use of cutting-edge technologies and adapting to evolving consumer preference, Amazon established itself as an innovative and flexible firm, ultimately changing the direction of cloud computing, e-commerce as well as a variety of other sectors.

Chapter 6: A Company Transformed

Amazon's Evolution from Online Bookstore to E-commerce Giant

Amazon's transformation from a modest online store to an international online retailer is an incredible tale of creativity, innovation and shrewd leadership. The journey of the company demonstrates the company's ability to spot developments, transform industries and constantly rethink its purpose to provide customers with the best service. The Amazon transformation story is as follows:

1. The Birth of Amazon:

Amazon.com was established by Jeff Bezos in 1994 as an online store. Bezos saw the potential for internet technology to transform the way we shop and started by providing an extensive collection of books which could be ordered easily and then delivered directly to the customers' doorsteps.

2. Diversification of Product Categories:

Following the successes of its book sales, Amazon quickly expanded into different products. Its determination to develop and expand its product offerings enabled it to be a single-stop source for an array of products for consumers, ranging such as clothing and electronics, to furniture.

3. Amazon Web Services (AWS):

In the year 2006 Amazon introduced a revolutionary change with the launch of Amazon Web Services (AWS) which is its cloud computing division. AWS offered businesses with scalable storage and computing capabilities and other solutions. The diversification to cloud computing was not just a source of significant revenue, but also highlighted the technological capabilities of Amazon.

4. Innovation in Customer Experience:

Amazon always concentrated on improving its user experience. They introduced services like personal recommendation, single-click ordering and reviews from customers that all increased customer engagement and loyalty.

5. Amazon Prime and Subscription Services:

The advent of Amazon Prime in 2005 revolutionized shopping online. Prime provided subscribers with fast and free shipping on all eligible products, as well as access to streaming media, among additional advantages. The subscription model encouraged customers' loyalty, and encouraged regular purchases.

6. Kindle and Digital Content:

Amazon changed the face of publishing through the launch of the Kindle ebook reader in 2007. The device let customers purchase and download digital publications which established Amazon as an important actor in the online content market.

7. Focus on Third-Party Sellers:

Amazon's third-party marketplace allowed independent sellers to market their items on

Amazon. The strategy diversified product offerings as well as increased the selection. It also let Amazon to generate revenue through fees and commissions.

8. International Expansion:

Amazon's popularity within its success in the United States prompted its international expansion. Amazon has established localized sites and products in a variety of nations and tailored its offerings to suit the diverse needs of different cultures as well as market trends.

9. Acquisitions and Innovations:

Amazon's acquisitions of strategic importance, including Zappos as well as Whole Foods, and innovations such as drone deliveries and cashier-less stores, have demonstrated its determination to stay in the forefront of business developments.

10. Amazon as an Ecosystem:

As time passed, Amazon evolved beyond e-commerce into an entire ecosystem comprised of services interconnected. From its platform for retail and cloud computing division, to entertainment services such as Prime Video and Alexa-enabled devices, Amazon established a vast digital footprint.

The evolution of Amazon from a bookstore online into an online giant is an example of its capacity to change as well as innovate and change industries. Amazon's focus on customer-centricity as well as its eagerness to adopt technological advances, as well as its revolutionary leadership of Jeff Bezos have propelled it into becoming one of the top and well-known companies globally.

Introduce of services such as Prime and Marketplace

Amazon's launch of products like Amazon Prime as well as the third-party Marketplace was pivotal to the development of its business and in establishing Amazon as a major player in the e-commerce and technology sectors. The services changed how people shop and communicate with Amazon, while expanding Amazon's revenue streams as well as its ecosystem. Let's look at the ways these services have transformed Amazon's operations:

1. Amazon Prime:

Amazon Prime, launched in 2005 it is a subscription service which offers a variety of advantages to its customers. The service revolutionized the online shopping environment and greatly increased the loyalty of customers. This is the way that Amazon Prime impacted the company:

Fast and Free Shipping: Amazon Prime members enjoy fast and free delivery on eligible merchandise. This option speeds up delivery of the item and encouraged buyers to select Amazon when they shop.

Streaming Media Prime provides access to Amazon Prime Video, a streaming service that offers the latest original content, movies as well as TV shows. The entertainment option added worth to the subscription as well as improved customer satisfaction.

Exclusive Discounts: Amazon often offers exclusive offers and discounts for Prime customers, helping users to buy more and generating a sense exclusiveness.

Prime Day: Prime Day is Amazon's yearly Prime Day celebration offers discounts and offers exclusively to Prime members, which creates the perfect shopping environment and increasing sales.

Customer loyalty: Amazon Prime's features build a solid bond between users and the service which makes it more likely for them to use Amazon for all their online shopping requirements.

2. Third-Party Marketplace:

The advent of the Third-Party Marketplace in 2000 permitted independent sellers to sell their merchandise on Amazon's Marketplace and significantly broaden the variety of products available. This change had many major effects:

Increased Variety of Products The Marketplace allowed Amazon to expand its variety of items without needing to keep track of inventory for each.

New Revenue Sources: Amazon makes money by commissions, referral fee as well as other fees by third-party sellers. This contributes to its overall profit.

Attracting Sellers: The Marketplace has attracted a wide range of sellers from smaller firms to well-known brand names, further improving your Amazon buying experience.

Fulfillment through Amazon (FBA) Amazon's FBA program permitted third-party sellers to use Amazon's extensive fulfillment network for packaging, storage and even shipping, improving the overall experience for customers.

Chapter 7: The Disruptor's Mindset

Amazon's Bold Moves: Acquisitions, Investments, and Risks

The success of Amazon can be traced partly to its disruptive attitude -- a desire to take risky actions, take strategically-planned acquisitions and investments and take risk-taking that is calculated. Jeff Bezos and his team are consistently demonstrating a forward-thinking strategy that has allowed Amazon to develop new ideas, increase its capabilities, and even shape the way industries operate. The Amazon disruptor's mentality has fueled its bold actions:

1. Acquisitions:

The strategic purchases of Amazon have had a key contribution to expanding its market and broadening its offering. A few notable acquisitions include:

Zappos (2009) acquisition of Zappos the online shoe retailer well-known for its excellent customer service, bolstered

Amazon's emphasis on providing outstanding customer service.

Whole Foods Marketplace (2017) Acquisition from Whole Foods provided Amazon with a retail location and access to the supermarket market and the potential for creating new supply chains as well as delivery.

Twitch (2014) The acquisition by Amazon of Twitch the live streaming service that caters to gamers, showed its commitment to the booming internet-based entertainment sector.

2. Investments and Innovations:

Amazon always invests in new ideas which allows it to remain one step ahead of its competitors:

AmazonWebServices (AWS): AWS revolutionized cloud computing by giving Amazon with a brand new income stream, and also changing the ways businesses make use of technology.

Amazon Echo and Alexa The creation of voice-controlled devices as well as virtual assistants have widened Amazon's product line and created innovative ways for users to communicate with tech.

"Drone Delivery: Amazon's study of drone delivery demonstrates its determination to redefine logistics, and creating more effective delivery solutions.

3. Calculated Risks:

Amazon's willingness to assess risks has resulted in important successes.

Prime Membership: Launching Amazon Prime with free, speedy shipping was a daring step that revolutionized customer expectations within the world of e-commerce.

Original Content (Prime Video): Amazon ventured into creating original content and entered the industry of entertainment, and compete against established companies.

Cash-Back Programs: The cash-back incentives and offers enticed users to utilize Amazon's payment service and buy more often.

4. Long-Term Focus:

Amazon's disruptionist mindset is set by a long-term vision:

"Reinvestment of Profits": Rather instead of prioritizing profits for the short term, Amazon often reinvests earnings into innovation and growth strategies.

Customer-centric approach: Amazon's primary focus on providing customers with value is in line with the long-term aim to build trust and create loyalty.

5. Continuous Adaptation:

Amazon's philosophy is constantly adapting to market dynamics that change:

Physical Retail Innovation: In spite of being a major online retailer, Amazon has ventured into physical retail stores such as Amazon Go stores as well as Amazon Books and is demonstrating its adaptability.

International Expansion: Amazon's expansion into international markets shows the company's ability to adjust to different cultures and business contexts.

It is clear that Amazon's disruptive approach is founded on creativity, risk-taking with a calculated stance, as well as a dedication to sustainable growth. Amazon's daring investment, acquisitions and determination to reinvent industries are establishing it as an industry leader in the field of e-commerce, entertainment, technology, and beyond.

Revolutionizing Industries: Cloud Computing and Beyond

Amazon's role in revolutionizing industries goes beyond e-commerce into fields like cloud computing and technology. One of the biggest achievements Amazon has accomplished is in the area of cloud computing, through the Amazon Web

Services (AWS) company. Furthermore, its innovative approach has brought about improvements across other industries. This is the ways that Amazon is changing the industries it serves:

1. Cloud Computing with Amazon Web Services (AWS):

AWS which was introduced in was a paradigm shifter within the tech industry. The company introduced the idea of cloud computing. It allows companies access to scalable and affordable computing capabilities without the requirement to build a large infrastructure. AWS can provide services that include storage capacity, computing power data, databases, analytics machine learning and much more. A few of the methods AWS has changed the landscape of technology are:

Scalability: Businesses are able to scale their resource in accordance with requirements, eliminating the need to make an upfront purchase of equipment.

Cost-Efficiency: The pay-as you-go model permits businesses to only pay for resources they utilize which reduces capital expenses.

Global Reach: AWS's global network of data centers allows firms to reach a global public with very low latency.

Innovation Acceleration by delegating administration of the infrastructure to AWS businesses can concentrate on creating innovative apps and services.

2. Voice Technology through Alexa:

The development by Amazon of Alexa as well as the Echo devices has brought the voice-controlled tech into the homes of people. Alexa's abilities range from basic voice commands, to the control of smartphones, offering information on weather conditions, scheduling reminders and carrying out tasks such as shopping, or listening to music. The technology has changed the way users use devices as well as accessing data.

3. The Entertainment Industry and Prime Video:

Amazon Prime Video disrupted the media industry through the creation of original content, and is challenging the traditional media outlets. With its investment in original films or TV series, as well as streaming media, Amazon has become a prominent actor in the world of entertainment with a broad selection of entertainment for its Prime customers.

4. Supply Chain and Logistics Innovation:

Amazon's efforts to optimize logistics and supply chain processes sets new standards for speed and efficiency within the business. The latest innovations like robotic fulfillment centers autonomous delivery vehicles as well as the development of drone delivery have revolutionized the way that products are packed, stored and shipped to customers.

5. Retail and Brick-and-Mortar Innovations:

Although it is mainly known for its online retail however, its expansion into brick and mortar ideas such as Amazon Go stores (cashier-less shopping) as well as Amazon Books (physical bookstores) illustrates the ability of Amazon to mix both traditional and technology-based retailing in new ways.

It is clear that the influence of Amazon goes beyond its online-based e-commerce origins. Amazon's innovations in the cloud, technology for voice entertainment, supply chain innovations and experiences for retail have transformed industries and created new standards in innovation as well as customer experience. Amazon's determination to challenge and redefine the norms of business is a major factor in the advancement of technology which continue to affect the way that individuals and businesses are able to operate in the digital age.

Chapter 8: Customer Obsession And Innovation Building A Culture Of Customer-Centricity

The success of Amazon is firmly rooted in its unwavering dedication to innovation and customer focus. Jeff Bezos and his team have created a work environment that puts customers in the forefront of all they are doing. The customer-focused approach has shaped Amazon's strategy, choices and innovation, influencing Amazon's reputation and growth. Let's look at the way Amazon established a culture of innovative thinking and customer-centricity:

1. Customer Obsession as a Core Value:

Since its beginning, Amazon established customer obsession as the primary value of its business. Its aim is to become the world's most customer-focused company. the company's commitment to customer service is evident in every part of the company's operations.

2. Data-Driven Insights:

Amazon makes use of data to study customer behaviour patterns, preferences and trends. Amazon uses insights derived from data to tailor recommendations and optimize customer experiences, and constantly enhance its services.

3. Long-Term Customer Relationships:

Amazon puts the importance of long-term customer relations in preference to short-term profit. This philosophy can be seen in programs like Amazon Prime with its emphasis is on providing benefits to the customers over time. This leads to loyalty and repeat sales.

4. Continuous Improvement:

Amazon is a company that encourages constant improvements. It aims to anticipate needs of its customers and tackle issues proactively. Feedback loops are crucial for finding areas in which Amazon is able to improve its service.

5. Innovating on Behalf of Customers:

Amazon's innovative mindset is founded in solving the problems of customers. It doesn't matter if it's improving the speed of delivery or introducing new products such as the Kindle or creating services such as AWS the company's innovations are created to enhance customer satisfaction.

6. Iterative Approach:

Amazon adopts a method of iteration to innovating. It isn't afraid of try new things and to learn from mistakes. The company's approach is to encourage employees to try new things and to learn from successes as well as setbacks.

7. Customer Feedback and Reviews:

Reviews and ratings from Amazon's customers give valuable insight to customers and Amazon. The transparency of reviews and ratings empowers consumers and assists Amazon improve its products in line with real-world observations.

8. Customer Service Excellence:

The Amazon customer service team is well-known for its effectiveness and efficiency. Amazon's focus is on speedy problem resolution, easy returns and placing the needs of its customers prior to all other requirements.

9. The "Two-Pizza Team" Approach:

Amazon uses a decentralized system that is known as"the "two-pizza team" approach. Teams are kept as small that they are able to eat with two pizzas. This encourages the ability to make decisions independently and with greater flexibility.

10. Leadership by Example:

Jeff Bezos himself leads by his example, frequently highlighting the importance of customer focus as well as innovation. The willingness of Jeff Bezos to risk his business and to invest in the long-term future is a reflection of the values that are reflected in the company.

Overall, Amazon's focus on its customers' approach is at the heart of its growth. Amazon's dedication to continual improvements, data-driven insight as well as innovations to meet demands of customers has allowed the company to change sectors, create new standards and offer value to its customers. The culture of obsession with customers and constant innovation is the main reason why Amazon stands out Amazon and has helped to earn its standing as a global market leader.

Innovative products such as Kindle, Echo, and Alexa

Amazon's inventions like Alexa, the Kindle, Echo, and Alexa are revolutionary and have changed how individuals use technology, consume information as well as perform daily tasks. The devices have not only proved the technical expertise of Amazon, but they have also become fundamental elements of the lives of many. We'll look at each one of them:

1. Kindle:

The Kindle launched in 2007 revolutionized the ways users read and browse books. It was a major factor in the popularity of e-books and ebook readers. Some of the key features and benefits of the Kindle are:

Accessibility to E-Books: The Kindle offered a simple method of carrying and gaining access to the entire library of books electronically, that caters to users their mobility needs and preferences for variety.

"E-Ink" Technology. application of E-Ink technology has made it possible to read on Kindle screen feel like writing on paper, which reduced the strain on your eyes.

Global Impact: Kindle allowed readers to purchase and download eBooks via wireless, making them accessible all users in the world.

Self-Publishing and Indie Authors Kindle Direct Publishing (KDP) let authors self-publish digitally their work making publishing more accessible to everyone.

2. Echo and Alexa:

Amazon's venture into voice-controlled technology using the Echo device as well as the Alexa voice assistant is an important step forward in computer-human interaction. Echo and Alexa's impact is Echo and Alexa's effects include:

Voice-activated assistance: Alexa became a virtual assistant that can answer queries, establishing alarms, playing music as well as controlling devices via the voice.

Smart Home Integration (Smart Home): Alexa's connection with smart devices in the home allowed users to control lighting or thermostats by using the voice of Alexa.

Skills and third-party integrations: Amazon opened up Alexa to third-party programmers, resulting in a wide range with "skills" that expand its capabilities.

Chapter 9: Chasing Moon Shots

The journey into Space Blue Origin's Genesis

"Chasing Moonshots" is a phrase that reflects Amazon's and Jeff Bezos' drive to take on exciting and revolutionary initiatives. An instance of this could be Blue Origin, Bezos' aerospace firm that is focused on exploration of space and technological development. Blue Origin's goal is to help create a space that allows millions of people to reside and work from space. Let's take a look at Blue Origin's history and mission. Blue Origin:

Genesis of Blue Origin:

Blue Origin was founded in 2000 by Jeff Bezos with the goal to make space travel more accessible and cost-effective. Blue Origin's name "Blue Origin" is a reference towards Earth in the sense of being it is the "blue origin" of humanity which represents the vision of the company of expanding the human population far beyond the planet.

Goals and Initiatives:

Blue Origin's goal is to develop technologies to enable secure and reliable space travel. Blue Origin's goals and initiatives are as follows:

1. Suborbital Flights:

Blue Origin's New Shepard vehicle is designed for suborbital flights. It will carry passengers and payloads all the way to the outer limits of space, allow them to enjoy just a few seconds of weightlessness and to observe the curvature the Earth.

2. Reusable Rockets:

The most important aspect of Blue Origin's strategy is to develop recyclable rocket technologies. This New Shepard rocket and capsule can be used to fly multiple times, thus reducing the expense of accessing space.

3. New Glenn Rocket:

Blue Origin is working on the New Glenn rocket, a high-lift launch vehicle that is designed to transport both uncrewed and crewed payloads in different orbits. New Glenn aims to compete in the satellite launch commercial market.

4. Lunar Ambitions:

Blue Origin has expressed interest in exploring the moon and is involved with NASA's Artemis program. The goal of the program is to bring humans back on the Moon. Blue Origin is currently developing technology to support lunar missions.

5. Blue Moon Lunar Lander:

Blue Origin unveiled its Blue Moon lunar lander design which is intended to move people and cargo to the lunar surface. It is a bid to aid in NASA's mission of creating an ongoing presence at the Moon.

6. Human Spaceflight:

The Blue Origin's New Shepard suborbital flights have not been manned as of yet The company is planning to transport paying passengers on its future flights providing a glimpse of space travel for private people.

7. Long-Term Vision:

Blue Origin's goal is to build a world that is open to space and humankind can exist and thrive far beyond the confines of Earth. Blue Origin believes that reducing the price of access to space is vital to achieve this objective.

Jeff Bezos has expressed his love for space exploration and his confidence in the potential Blue Origin's potential Blue Origin to shape the human future. Blue Origin's strategy of developing technologies incrementally as well as its dedication to reuse are in line with Bezos his long-term goals and commitment to Moonshot goals.

To conclude, Blue Origin exemplifies Amazon's as well as Jeff Bezos' penchant for "chasing moonshots." By advancing innovative technology and determination to achieve ambitious objectives, Blue Origin is contributing to the growth of space exploration, and promoting an era where space becomes much more accessible to all.

The Long-Term Vision for Space Exploration

Jeff Bezos and Blue Origin are articulating a compelling, longer-term plan for space exploration that is more than immediate goals, and envisions a world where humans have a continuous and flourishing presence within space. This vision is marked by sustainability, accessibility, as well as the potential of humanity to extend its presence over Earth. Below are a few of the most important aspects of this vision for the future:

1. Ongoing Access to Space:

Blue Origin's mission is the making space travel available to a wider range of individuals. It's not just scientists and astronauts, but too tourists, researchers as well as people who are interested in exploring space.

2. Lowering the Cost of Space Travel:

The main focus of Blue Origin's strategy is reducing the expense of traveling to space. Blue Origin's emphasis on reuse which is evident in vehicles such as New Shepard and New Glenn that are designed to fly multiple times, is designed to cut the cost of launch down and giving greater use of space.

3. Building Infrastructure:

Blue Origin envisions creating the essential infrastructure for space in order for long-term human settlement. This is a matter of technology that supports transportation for habitation, habitat, utilization of resources and much more.

4. Space Tourism:

A key aspect of the plan is offering space travel opportunities for civilians, which allows them to explore space and feel weightlessness as well as the overall impact - the profound change in perception that results from being able to see Earth in space.

5. Lunar Exploration:

Blue Origin's lunar goals are in line to NASA's Artemis program. The goal of the program is for humans to be returned in the Moon. Its vision includes the development of lunar landing craft and other technologies that will aid in lunar exploration as well as the possibility of habitability.

6. Colonization of Space:

Bezos has talked about the possibility of shifting the heavy industry and pollutants away from Earth and in space, and possibly saving the planet and expanding humanity's activities and civilization to spaces.

7. Creating a Future for Millions in Space:

The final aim is to provide the future in which millions of people work and live in space. The vision of space is an area that is not only to be explored, but as a place that can create possibilities for new industries, opportunities as well as communities.

8. Space as a Resource Frontier:

Blue Origin's vision of space sees it as a resource source as well as energy and innovating. Exploring asteroids to extract precious minerals as well as harnessing solar energy within space are some of the options.

9. Collaboration and Partnerships:

Blue Origin's strategy involves cooperation with various organizations, such as public and private companies as well as international organizations. Partnerships are considered essential for achieving the objectives in space research.

Chapter 10: Challenges And Criticisms

Labor Practices and Worker Conditions

Although Amazon and its CEO Jeff Bezos have achieved remarkable successes and breakthroughs however, they've faced various problems and criticisms. The most significant issue is the way in which Amazon conducts its labor practices as well as workplace conditions that affect employees within the operation. The issues have led to discussions concerning employee conditions as well as workplace culture and the effect of Amazon's business practices regarding its workers. Below is a brief overview of the issues and concerns regarding labor policies as well as working conditions.

1. Working Conditions:

Some critics have expressed concerns over working conditions at Amazon's fulfillment centres, where workers are responsible for selecting, packing and shipping packages. The reports have revealed intense physical requirements, long working days, and short breaks. The aforementioned conditions have caused discussions regarding the health of workers and their well-being.

2. Employee Burnout:

The fast-paced atmosphere at Amazon and its focus on efficiency have come under fire as a factor in employee burning out. Workplace pressure and high expectations of productivity could cause stress and fatigue within employees.

3. Injuries and Safety:

The reports of injuries at work at Amazon's fulfillment facilities are raising questions about the safety of workers in their workplace. The risk of injuries and strains resulting from working in a physical environment work are causing discussions on how to ensure the health and safety of employees.

4. Surveillance and Monitoring:

According to some reports, Amazon utilizes surveillance and monitoring equipment for tracking employee productivity as well as movements. The company has been criticized for how to balance efficiency with the privacy of employees.

5. Unionization Efforts:

There have been attempts from a few Amazon employees and groups to form unions and organize. Amazon has been criticized over its attitude towards unions, as well as accusations of anti-union strategies.

6. Response to Criticisms:

Amazon is being criticized over the way it has dealt with labor issues. Some critics say that Amazon's response has not been adequate or even dismissive regarding the concerns brought up.

7. Repercussions on Brand Image:

Labor working conditions and practices could affect the image of Amazon's brand because public consciousness of these issues increases and the public's perception of these issues changes.

8. Wage and Benefits Discussions:

The Amazon debate has erupted over its pay and benefits offered to its employees. Many argue that Amazon ought to increase its wages and offer more benefits in order to meet problems faced by employees.

9. Ongoing Scrutiny:

Amazon's labour practices have received significant attention from the media and have drawn the public's attention and scrutiny. This has led to growing demands for accountability, transparency as well as improvements to the way workers are treated.

It is important to know it's important to note that Amazon has taken action to tackle a few of these worries. Amazon has announced plans that will improve safety in the workplace and provide benefits to employees and provide employee training and training. However, the concerns and concerns about labor policies highlight the conflict between the needs of an ever-growing online retailer and the health of its workers. Discussions about these topics remain a major focus of discussion regarding Amazon's effect on its workers and on the overall economic system.

Balance between Innovation and Social Responsibility

The balance between innovation and ethical and social responsibility can be a challenging issue faced by many businesses which includes Amazon. Being a pioneer in the field of the field of technology and commerce, Amazon has a significant influence on many elements of our society, ranging from working practices to the economy as well as sustainable environmental practices. Finding a way to balance cutting-edge technology and preserving social responsibility requires a careful evaluation of both business and ethical issues. This is the way Amazon as well as other businesses will be able to navigate this dilemma:

1. Ethical Leadership:

Leadership is a key factor in determining the tone of an organization's strategy for innovative thinking and social accountability. Decision-makers and executives must consider ethical concerns and coordinate their business plans with the wellbeing of their employees, customers and their communities.

2. Transparent Practices:

Transparency in business operations can help increase trust among all the stakeholders. Businesses should be transparent about their innovations as well as social responsibility initiatives and their progress in dealing with challenges. Transparency helps to improve accountability and promotes constant improvements.

3. Employee Well-Being:

Making sure that employees are well-being a priority is crucial. Employers can put their money into decent wages, secure working conditions, balance between work and life and opportunities for professional growth. A happy and enthusiastic workforce is a key ingredient in innovations and positive social impacts.

4. Sustainable Innovation:

Innovation shouldn't come without sustainable development for the environment. Businesses can design and develop strategies and techniques that help reduce their carbon footprint, cut down on consumption, and increase sustainability in resource usage.

5. Community Engagement:

Connecting with communities in the local area and understanding their requirements can assist companies to tailor their innovation to make positive impact. Communities and partnerships are a great way to tackle social issues and drive the development of new products and services.

6. Responsible Supply Chains:

The company should make sure that its suppliers adhere to the highest standards of ethics and sustainability. In encouraging fair labor practices and ethical sourcing, businesses can create new ideas while also making a contribution to the environmental and social advancement.

7. Social Impact Assessments:

Making impact assessments helps businesses understand the possible consequences of their ideas on a variety of stakeholder groups. It allows them to take well-informed decisions that take into account both the social and technological aspects of innovation.

8. Philanthropy and Giving Back:

The companies can utilize their funds to fund social causes and help communities in need. It could be through philanthropic projects as well as charitable contributions and programs to address social problems.

9. Collaboration and Partnerships:

Working with government agencies, non-profit organizations academics, universities, as well as other organisations will result in more integrated methods of innovation that take into consideration the social responsibilities aspect.

10. Long-Term Perspective:

Although innovation is often focused on the short-term, businesses must also think about the long-term impacts of their innovations on the society. It is important to consider the potential risk and the benefits that might be realized over the course of the course of.

Chapter 11: Personal Life And Philanthropy

The Private Side of Bezos: Family and Relationships

Jeff Bezos, while widely recognized as the originator of Amazon as well as a tech-savvy businessman, is also very private about his personal life, which includes his relationship with his children and family. What is known about his personal life:

Marriage and Divorce:

Jeff Bezos was married to MacKenzie Scott (formerly MacKenzie Bezos) from 1993 until in the year. They met when they worked together for the firm that managed investments D.E. Shaw during the 1990s. The couple got married in 1993, which was a year prior to when Jeff established Amazon. In their wedding they had four kids in their lives.

In January of 2019, Jeff and MacKenzie announced they would divorce following some time of separation. The divorce was completed in April of 2019. In the agreement to divorce, MacKenzie received a significant part of Amazon's stock, which makes her among the most wealthy people in the world.

Children:

Jeff Bezos and MacKenzie Bezos have four children with them: three sons and a daughter. Their private lives have largely remain private. Very few details are publicly accessible regarding their children.

Relations and Lauren Sanchez:

After his separation from MacKenzie There were rumors regarding Jeff Bezos' relationship with Lauren Sanchez, a former anchor of a TV show and an upcoming media star. Jeff and Lauren's romantic relationship was made known in the early part of 2019 at the time that the divorce announcement was made. Media coverage of their marriage attracted significant interest.

Privacy and Personal Life:

Jeff Bezos is known for his private life in his private life. The Amazon founder has tended to keep his personal relationships and family off the radar and has focused exclusively on his job in Amazon as well as other businesses. The privacy he has enjoyed allows him to keep a distance from his private life and his work as a well-known business person.

It is important to remember that my information comes from information that was accessible up to September 2021. There could have been changes in Jeff Bezos' personal life in the years since that time, which aren't included in this report. Bezos's focus on preserving his privacy has resulted in a lack of information available of his relationships with family members and concerns regarding his family.

The Bezos Day One Fund and Climate Initiatives

The Bezos Day One Fund and Jeff Bezos' climate initiatives demonstrate his commitment to charity and taking on significant global challenges, especially the ones relating to education and climate changes.

1. Bezos Day One Fund:

The Bezos Day One Fund was created in the name of Jeff Bezos and his then-wife, MacKenzie Scott, in September, 2018. The name of the fund comes from Bezos his belief that each activity, whether it's professional or personal, should be approached like it was "Day One." The fund's structure allows it to concentrate on two primary aspects:

The Day 1 Families Fund: The fund's goal is to offer help and assistance to those that work to combat homelessness and aid families in need of financial assistance.

The Day 1 Academies Fund: Its primary goal is the creation of a network of premium Montessori-inspired, preschools that are accessible to under-served communities.

Through these projects, Bezos aims to contribute positive social changes as well as provide opportunities for people as well as families that may have difficulties in accessing resources and education.

2. Climate Initiatives:

In February of 2020, Jeff Bezos announced the creation of the Bezos Earth Fund, committing $10 billion in order to combat climate change and encourage sustainable development of the environment. The goal of the fund is to help research, organizations activist groups, initiatives, and other organizations which are working towards tackling the effects of climate change, create sustainable energy options and safeguard the earth's natural resources.

Bezos is adamant that climate change is the most significant threat that humanity confronts, and believes the efforts of both individuals and groups can be crucial to tackling this issue. Bezos Earth Fund Bezos Earth Fund aims to offer resources that will help accelerate the pace of innovation and promote sustainable practices and help create a better and green future for the globe.

The Bezos Day One Fund and the Bezos Earth Fund exemplify Jeff Bezos determination to leverage his power and wealth to effect positive transformation. The philanthropic efforts focus on solving pressing environmental and social problems and reflect his determination to impact the world beyond the role of an executive in the business world.

Chapter 12: The Washington Post And More

The Acquisition of The Washington Post

In August 2013 Jeff Bezos made headlines by purchasing The Washington Post, a well-known and reputable newspaper that has long history in American journalism. The purchase was made by Bezos personaly, independent of his work in Amazon which was an important step into the world of media. Below are the particulars as well as the implications of Bezos purchase of The Washington Post:

1. Acquisition Details:

Jeff Bezos purchased The Washington Post for $250 million cash. He bought the paper from its former owner The Washington Post Company (now named Graham Holdings Company). This acquisition was completed as a private investment made by Bezos and wasn't tied to Amazon's business activities.

2. Personal Investment:

Bezos explained that his purchase of The Washington Post was a private investment, and that it would be run independently of Amazon. Bezos emphasized his dedication to ensuring the newspaper's editorial independence, and he also praised its journalism efforts.

3. Preservation of Editorial Independence:

One of the most important features of the deal was Bezos his commitment to protecting the Washington Post's journalistic independence and editorial independence. He promised that he wouldn't interfere with the reporting of the newspaper or editorial choices.

4. Digital Transformation:

Bezos has brought his business-minded and technologically-savvy style into The Washington Post. As part of his management the paper underwent major digital changes, with a focus on cutting-edge methods of digital journalism as well as strategies to change with the evolving media world.

5. Investments in Technology and Talent:

Bezos made investments in upgrades to technology and other initiatives that will enhance the Washington Post's web presence and its digital offerings. Additionally, he helped in the hiring of the best journalists as well as technology to fuel the growth of the newspaper and to foster innovation.

6. Financial Stability:

Bezos his ownership provided the stability of finances the stability of The Washington Post, enabling the newspaper to invest in investigative journalism, as well as other quality content, even when the conventional media industry struggled.

7. Innovations and Expansions:

Under Bezos under his control, The Washington Post implemented numerous innovations, such as media-driven journalism that relies on data along with multimedia storytelling and increasing its subscription options for digital. The goal was to attract readers, and to adapt to changing trends in consumer behavior.

8. Contributions to the Media Landscape:

Bezos the purchase the newspaper The Washington Post highlighted his desire to support independent journalism as well as helping to create a dynamic media ecosystem. Personal involvement in the acquisition also highlighted that there is a possibility of cross-industry collaborations between technology leaders and traditional media.

In short, Jeff Bezos' acquisition of The Washington Post showcased his fascination with journalism, the media and the significance of a democratic press within the world. The Post's ownership provided financial stability along with innovation and a online-first focus for the paper but also preserved the newspaper's freedom of editorial. Bezos's involvement in the world of media was not limited to his position as CEO of Amazon and demonstrated his wider influence and dedication to a variety of industries.

Bezos' Role in Journalism and Media

Jeff Bezos' role in media and journalism has been noted not just for the ownership he has of The Washington Post but also for his contribution to the transformation of the media industry to digital as well as his commitment to promoting the development of high-quality journalism. These are the most important aspects of Bezos influence on media and journalism

1. Digital Innovation:

Bezos has brought his tech-savvy perspective his tech-savvy approach to The Washington Post, encouraging The Post to adopt the digital revolution. As part of his management The newspaper invested in technological upgrades in data-driven journalism as well as multimedia storytelling, to be able to adjust to the evolving media landscape and to engage its readers in innovative ways.

2. Emphasis on Quality Journalism:

Bezos has repeatedly stressed the importance of a high-quality journalistic work as the foundation for a strong democratic society. Bezos has been a strong supporter of the Washington Post's investigative journalism and initiatives to create deep, compelling journalism.

3. Business Model Experimentation:

With his control over The Washington Post, Bezos has tried out a range of ways to support journalism into the new digital era. These include initiatives that aim to boost digital subscriptions as well as explore other ways to generate revenue that go beyond the traditional model of advertising.

4. Cross-Industry Collaboration:

Bezos The owner of The Washington Post has demonstrated that there is potential for collaboration across industries between tech leaders and traditional media. The involvement of the founder has highlighted the need for creative solutions to meet the issues facing journalists.

5. Philanthropic Support:

As well as his ownership of The Washington Post, Bezos has also launched his own Bezos Earth Fund by pledging $10 billion for issues related to climate change as well as environmental sustainability. Although it isn't directly connected to journalism, the initiative is a reflection of his determination to use his assets to tackle the global issues.

6. Media Landscape Influence:

Bezos his position as an acclaimed tech entrepreneur and media owner has provided him with the opportunity to have a significant impact on debates about the importance media plays in the society and the effects of technology advancements, and the issues that traditional news organizations face.

7. Balance of Business and Editorial Independence:

In his time as the owner of The Washington Post, Bezos is determined to keep an equilibrium between his commercial interest and the paper's editorial independent. He has reiterated his pledge to not interfering in the reporting and decision-making process.

8. Conversation on Press Freedom:

Bezos's involvement in the ownership of media has provoked discussions on the importance of press freedom, influence of media owned by wealthy people as well as the significance of securing independent editorial integrity.

Chapter 13: Wealth, Influence, And Legacy

Bezos' Rise on the Forbes Billionaires List

Jeff Bezos' rise on the Forbes Billionaires List is a evidence of his incredible achievement as an entrepreneur and to the rapid growth of Amazon as an international technological and e-commerce powerhouse. The journey of a start-up leader to one of the richest people in the world is a story of innovation as well as risk-taking and ever-growing business of Amazon. This is a brief outline of his rise to Forbes' Forbes Billionaires List:

1. Early Wealth Accumulation:

Jeff Bezos founded Amazon in 1994 with the intention of creating an online bookshop. Amazon's rapid expansion and its success during the initial times led to substantial earnings growth for the Bezos family as Amazon emerged as a leading market in the world of e-commerce.

2. Expanding Business Verticals:

When Amazon broadened its product offerings beyond its books and expanded into other products and services, the company's revenue and market capitalization increased significantly. Its expanding into new fields such as cloud computing (Amazon Web Services) as well as media (Amazon Prime Video) further helped in its financial success.

3. Stock Performance:

A large portion in Jeff Bezos' wealth is linked to his ownership of the Amazon stock. The price of Amazon's stock has seen significant increases throughout the years, fueled by its strong financial performance in the market, its dominance over other companies, as well as investor confidence.

4. Continuous Innovation:

Bezos his reputation as a leader who is innovative and Amazon's constant pursuit of innovative ventures and new technologies are a major factor in the company's worth and wealth. The innovations such as Amazon

Prime, Kindle, and Alexa has helped keep the company's popularity and its profitability.

5. Personal Investments:

Bezos his personal investments, including the ownership of The Washington Post and his dedication to space exploration via Blue Origin, have also led to his rise in the public spotlight as well as his wealth.

6. Ongoing Global Expansion:

Amazon's global expansion to a variety of countries as well as its diversification into different sectors have enabled the company to explore new revenues and markets which has further increased Bezos riches.

7. Wealth Fluctuations:

Bezos's wealth has gone through changes due to things like stock market fluctuations, the financial performance of Amazon and shifts in the way consumers behave. Bezos' net worth is dependent on market dynamics as well as the economic outlook.

8. Philanthropic Initiatives:

Recently, Bezos has become increasingly engaged in charitable endeavors which include his Bezos Day One Fund and the Bezos Earth Fund. These initiatives require substantial investment in funds, they represent his intention to invest the wealth he has amassed to benefit environmental and social causes.

Jeff Bezos' rise on the Forbes Billionaires List showcases his ability to convert a start-up into one most successful companies, as well as his influence on the technology and e-commerce industries. The influence of Jeff Bezos extends far beyond the corporate, because his actions in innovation, investments, and innovations have had an influence on the society, media, as well as the ways consumers live and use goods and products and.

Shaping the Future: What Will Bezos Be Remembered For?

Jeff Bezos will likely be recognized for a variety of achievements and contributions that have transformed the technological world of business, and the entire society. The legacy of Jeff Bezos will span several important areas of his work:

1. E-Commerce Revolution:

Bezos the pioneer of his role in the development of Amazon and transforming the way customers shop online will form the mainstay of the legacy. The retail market was transformed by him and introduced the concept of e-commerce as well as transforming Amazon to become a world-class marketplace.

2. Technological Innovation:

Bezos his focus on technology and innovation has resulted in the development of services and products that are now a staple of our lives. Innovative products such as Amazon Prime, Kindle, and Alexa have established new standards in ease of use, entertainment, as well as communications.

3. Entrepreneurial Spirit:

One among the top entrepreneurs of the history of business, Bezos' journey from the beginning of a small-scale startup in his garage to becoming an industry giant will encourage future entrepreneurs to follow. His determination, vision and willingness to make a risk are a model for the business world's leaders.

4. Disruptive Mindset:

Bezos his ability to shake up traditional sectors, challenge the expectations, and predict the market's trends has had an indelible impression. The disruptive approach he has adopted is not limited to retail, but also has influenced other industries such as cloud computing and media as well as logistics.

5. Shaping the Internet Era:

Bezos' contribution to the expansion of the internet period cannot be overestimated. The work he has done in partnership with Amazon is not just contributing to the growth of the digital economy but has also changed the ways businesses function on the internet.

6. Space Exploration:

Bezos his involvement with Blue Origin and his dedication to space exploration might be a major chapter in his career. Bezos' efforts to increase the accessibility of space and support the advancement of space-related technologies will have a profound impact on the future of humanity.

Chapter 14: Transition And New Beginnings

Bezos' Decision to Step Down as Amazon's CEO

The 2nd of February, 2021 Jeff Bezos announced his decision to leave his position as the CEO of Amazon, marking the beginning of a major transition for Amazon as a company as well as Bezos himself. Below is the breakdown of the decision and its consequences it has:

1. Leadership Transition:

Bezos declared that he was going to change his position as Amazon's Chief Executive Officer to take on the post as the The Executive Chairman of Amazon's board. As Executive Chairman his focus would be on the strategic direction of Amazon, its most important projects and the longer-term plan for the business.

2. Andy Jassy as CEO:

In the course of the plan for transition, Andy Jassy, who had worked for Amazon from 1997 and was a key player in the development of Amazon Web Services (AWS) was appointed the company's new CEO. Jassy's in-depth knowledge of the Amazon's business operations as well as the role he played in the success of AWS was a perfect fit to be the chief executive.

3. Motivation for Transition:

In his statement, Bezos stated that he is making the shift to pursue his interests and other projects such as the space exploration business Blue Origin, The Washington Post as well as philanthropic endeavors. Bezos expressed his confidence that Jassy will be able to manage Amazon's day-today business.

4. Amazon's Growth and Impact:

Bezos made the decision at an era when Amazon was growing into one of the biggest and most influential corporations that spans multiple areas and industries. Under his guidance, Amazon had transformed e-commerce and pioneered cloud computing through AWS as well as disrupted conventional business methods.

5. Legacy and Future Vision:

Bezos said he was strongly dedicated to Amazon and its long-term achievement. The company was expected to continue to innovate and add the best value for its customers, while retaining its commitment to customer service.

6. Long-Term Vision:

Bezos the decision of stepping off as CEO was accordance with his vision for the future and his approach to the growth of Amazon. He was aware of the need for the strategic vision of his company and continuity in its leadership in the midst of an expansion phase. Amazon began to expand into new markets and faced new problems.

7. Personal Transition:

Bezos his change from the position of CEO to Executive Chairman enabled him to pursue projects and initiatives that went that were not related to Amazon and its interests in media, space exploration and the philanthropy sector. The transition marked a new phase for his professional life, which allowed Bezos to devote more of his time to the areas that in which he was passionate.

8. Impact on Amazon's Culture:

Bezos' style of leadership and principles made a huge impact on the culture of Amazon's company. He was a firm believer in customer obsession as well as innovation and long-term planning helped define the DNA of Amazon as well as the principles that are expected to remain in place under the leadership of Jassy's.

To summarize, Jeff Bezos' decision to quit as Amazon's chief executive was a major transition in the direction of one of the most powerful companies. This allowed him to focus on his other businesses as he entrusted the day-today business to his successor Andy Jassy. Bezos' exit reflected his vision for the future and dedication for the continued growth of Amazon.

Exploring New Ventures and Endeavors

Following his departure as the CEO of Amazon, Jeff Bezos has focused on pursuing new business initiatives and opportunities outside Amazon's core business. Transitioning from daily operations at Amazon allows him to devote his time and energy to initiatives that are in line with his interests and passions. Below are a few noteworthy new ventures and projects which Bezos is involved with:

1. Blue Origin:

One of Bezos major post-Amazon endeavors includes Blue Origin, his space exploration firm. Blue Origin is dedicated to creating technologies and vehicles that allow to travel through space, and the goal is opening space to everyone. Bezos' love of space exploration is the driving force behind Blue Origin's work to create reusable rockets that can be used for commercial space exploration.

2. The Washington Post:

Bezos remains the owner and continues to supervise The Washington Post, a well-known newspaper that he purchased in the year 2013. Even though he's been able to step away from running the day-to-day activities, the ownership of The Washington Post has enabled him to have an impact on the digital evolution of the newspaper and its place in the world of modern journalism.

3. Bezos Earth Fund:

In February of 2020, Bezos launched the Bezos Earth Fund by making a pledge of $10 billion in order to combat the issue of climate change as well as encourage environmental sustainability. It aims to help organisations and projects that are working in the fight against climate change as well as conserving our planet.

4. Philanthropy:

Bezos has been increasingly involved in charitable endeavors that go outside of his Bezos Earth Fund. Previous initiatives including his Bezos Day One Fund, focus on combating homeless, helping with education as well as addressing the social challenges.

5. Day 1 Fund:

The Day 1 Fund, established prior to Bezos change from the CEO position, is continuing to goal of addressing the issue of homelessness as well as early elementary education. The fund's creation reflects his determination to using his money to support educational and social initiatives.

6. Personal Projects and Interests:

Bezos change of direction has enabled him to pursue private initiatives, hobbies and hobbies that he would not have had the time to pursue while running Amazon. The personal pursuits may include creative endeavors as well as recreational activities.

7. Innovation and Investments:

Bezos is known for his ingenuity and creativity. the driving factor behind his work post-Amazon. The CEO is set to investigate different investment options and participate in initiatives that are aligned with his future vision.

As a summary, Jeff Bezos' transition as Amazon's chief executive is opening up new avenues for him to investigate an array of new projects and ventures. The focus he has placed on exploring space along with media ownership, charity, and many other interests highlights his determination to make an impact globally and shaping the future of different areas.

Chapter 15: Closing Thoughts

Jeff Bezos' Enduring Impact on Business, Technology, and Society

Jeff Bezos' enduring impact on technology, business as well as society has been unquestionably revolutionary. The journey he took from start-up founder with a small garage into an one-of-a-kind prominent individuals around the globe is etched in a myriad of fields. Below are some thoughts about his legacy

1. E-Commerce Revolution:

Bezos changed the way that people buy and transformed the shopping marketplace by launching Amazon. The vision of shopping online came to fruition, profoundly shifting consumer behavior as well as changing the way that e-commerce is conducted globally.

2. Digital Innovation:

Bezos His constant search for invention led to the development of services and products that are now part of everyday life. His innovations in digital reading using the Kindle Smart homes, smart devices that use Alexa and cloud computing using AWS have set new benchmarks.

3. Disruption and Adaptation:

Bezos his disruptive approach to business has challenged the established business strategies. The Amazon founder demonstrated the need for being flexible in a world technological change and changing markets in a way that encouraged companies to stay in the forefront of technology.

4. Customer-Centric Philosophy:

Bezos's obsession with customer service is now a characteristic of Amazon's growth. Bezos' commitment to providing satisfaction to his customers was the groundwork for building trust, loyalty and culture-based customer service in organizations across the globe.

5. Job Creation and Economic Impact:

The growth of Amazon under the leadership of Bezos has led to the creation of millions of jobs both directly and indirectly. They have significantly contributed to the world economy as well as employment possibilities.

6. Technological Advancements:

Bezos his innovations and investments in technology has pushed the boundaries of what is possible. The influence of his company extends far beyond Amazon to advancements in space travel, technology and sustainability for the environment.

7. Legacy of Entrepreneurship:

Bezos his entrepreneurial path is a guideline for young business executives. The ability of his to risk it all to think large, make big decisions, and implement his plan is an example for all those looking to establish and expand impactful businesses.

8. Philanthropy and Social Responsibility:

Bezos dedication to charitable giving and solving social issues reflects his understanding of the responsibility associated from his accomplishments. His commitment to climate change, education and other social causes show the desire to leave an impactful legacy.

9. Redefining Leadership:

Bezos his leadership style and philosophy has challenged the traditional concepts of leadership for business. The emphasis he places on thinking long-term flexibility, adaptability and a focus on the customer has created innovative leadership models for success.

Chapter 16: Modest Beginnings

It's a bit odd that billions or millions around the world know of Amazon.com is, but they know very little or no information regarding Jeff Bezos.

Jeff Bezos is an entrepreneur par excellence. He's a genius and was for a time the most wealthy person on earth. He is among the wealthiest people on earth. Bezos is hoping to become the world's first Trillionaire ever, and needs $910 billion in order to realize the vision! Jeff Bezos founded the number most popular online store in the world, which is generating $43.7 billion through the third quarter of 2017. in 2013, Bezos purchased Washington Post The largest publication across the United States, for $250 million.

Birth of an Icon

Jeff Bezos took his first breath on the 12th of January 1964, in Albuquerque the largest New Mexico city. After that, he was named Jeffrey Preston Jorgensen.

His father's biological father, Ted Jorgensen, married his teenage mother, Jacklyn Gise. Together, was he stayed for just one year. Ted worked in a modest market in the neighborhood and found it difficult to sustain his financial needs. He drank a lot and that proved to be the straw breaking his camel's back. She divorced the couple.

Jacklyn was introduced to Miguel Bezos, an immigrant from Cuba that came to Miami in 1962. She was married to him within six years. When Jeff was four years old, his Name was changed from Jeffrey Preston Bezos since Ted Bezos, his father's biological son, was not opposed to the adoption.

Childhood

In his early years, Jeff showed early snippets of wisdom and courage. Imagine a young child dismantling his crib with an electric screwdriver, as it was his goal to be in a bed with an adult. The shrewdness of this kid would assist Jeff build a fortune as well as lavishness.

Jeff was a young Jeff was born in an Catholic institution, and worked to improve his performance every day. He learned the English Language intently and did mundane jobs for hours so that to pay for his own studies through The University of New Mexico.

Jeff was the youngest of two brothers He had two brothers - Mark as well as Christina. When he was twelve, he'd already created robotics, made models, as well as various electronic kit. After he graduated from the high school level, Jeff literally converted their garage into an electronics and science lab. It was there that Jeff would launch Amazon.com many years more later!

The savvy business mogul became fascinated with technology. transformed their garage into an advanced lab while experimenting with electrical devices as well as appliances in the home. When he was a teenager, his family relocated from the city of Miami,

Florida, where the computer became his favorite.

Jeff gained a lot of attention in his new school, Palmetto High School, and would later be the class president and become a valedictorian.

The prodigy began his first venture into business: The Dream Institute- an instructional program for kids starting in fourth grade and up to the sixth grade.

Inspired by Grandpa

Since he was four years old, until the time he was 16years old, Jeff was a summer camper on the Texas ranch owned by his grandparents. In the end, the young man played with lots of farm equipment, and removed bulls' testicles!

One of Jeff's initial sources of inspiration was his grandfather, Preston Gise. He sparked in him his unwavering enthusiasm for creativity activities. When he delivered his commencement address, Bezos didn't forget to refer to his father's catchy quote, "It's more difficult to be kind than smarter."

In school, the teachers would be constantly amazed inspired by Jeff's amazing insight and wisdom. Jeff once told them that "Mankind's future is not on this planet." Jeff shared his hopes to become a space-based entrepreneurs. Being a man who aligns the words with action, Bezos currently owns Blue Originan organization that is involved in exploring space.

Jeff was a fan of sequels to the first version of Star Trek and even contemplated the use the name Amazon, MakeitSo.com, alluding to a well-known script written by one of the characters in the film Captain Jean-Luc Picard.

From Princeton to Wall Street

Bezos has attended Princeton University in New Jersey and was awarded the summa cum louise in the year 1986. He earned a bachelor's degree from the fields of Computer

Science (which was his primary area of study) as well as Electrical Engineering.

Following graduation, the talented young man knew exactly what was important to him and pursued the goal straight up. The young man received offers from numerous famous firms that are on Wall Street like Fitel or Furukawa Information Technologies and Telecommunications, Bankers Trust and the worldwide investment management company D.E. Shaw. Then, Jeff met his wife, Mackenzie, a famous American novelist hailing from California. He was the youngest vice-president of the business in just 4 years.

Image Obtained from: https://www.wired.com/2011/11/ff_bezos/

While working at an firm that managed investments when he worked at an investment management firm, he came across an article about the world wide internet, and it claimed that it was increasing by 2300 per year. A hunch he had told him was the case. He was certain he had to explore this enormous opportunity to trade, however he was not sure what kind of venture that could be effective. Jeff carried out a great deal of studies on mail-order companies He listed them as 20 over the course of time. He then picked books with huge potential for sales.

A champion for sure, Bezos made a risky move and entered the nascent realm of e-Commerce even though he had a lucrative and lucrative professional. Jeff decided to step down from his high-paying job in 1994. He then relocated to Seattle and began to explore potential of an unproven marketplace online by creating an online bookstore.

Fearing being a loser, David Shaw tried to convince Jeff to stay, but there was no way to stop Jeff, who was determined to succeed, from setting up in place his own business. Jeff would prefer to fail rather than not try to even try. This is the mindset of an accomplished person. Better to fail when exploring new areas than to remain in the comforts of mediocrity and ease. That's how our greatestness is birthed.

Jeff and Mackenzie traveled to Texas and rented a vehicle from their father prior to driving towards Seattle. While the couple was able to witness the stunning sunrise over the Grand Canyon, Bezos made certain calculations and projected the revenue. The result was the beginning of Amazon!

Chapter 17: Emergence Of An Empire

The 16th of July, 1995 will surely be the most important day of the career of Jeff Bezos. Bezos determined that it was the right the time to strike out independently. After a long and careful process of preparation the founder of Amazon.com was able to establish his own company and named it in honor of the mysterious Amazon River. Amazon.com's initial success was awe-inspiring. Without the hype of media it could sell books across the United States and 45 other nations in a month and produced an estimated $20, 000 from its revenues. It was more rapid then Jeff and his team of ragtag members had ever imagined. The team could never have anticipated the magnitude of their success. He defied every expectation.

It was a more sweet story in the two years following the commercial success, Amazon.com was incorporated. Naturally, there were many doubters among the experts in market research. What is the best way for a newbie to be able to compete through e-

Commerce sites? The answer was that they had a wrong answer - extremely incorrect. The year 1999 was the time when Amazon wasn't just in a position to compete, but also outperformed the competition, and became as the leader of the world of e-Commerce. Great, isn't it?

Bezos announced his IPO of shares at the price of $18 each. At the time Bezos was an established Amazon creator had a solid strategy in place, but the majority of investors were worried about the future of the business since it was not planning to earn a profit during the initial five years. In reality, it proved to be a blessing since Amazon was able to survive its initial phases, managed to escape the dot.com bubble, or market crash that occurred between 1997 and 2001. Others technology companies were not as fortunate, and they were a victims of the infamous financial crash.

The initial three months of 2001 saw more profits for Amazon.com. Amazon.com raked in revenues of $1 billion with a profits of $5 million. While the gain was rather small, it was a sign how Jeff Bezos was right all throughout. It proved he observed trends with precision. He proved he was a visionary.

No one could had imagined that Amazon would become the top online retailer. Looking back, suggesting that it was unjustified wouldn't be inappropriate. It's rather surprising, but surprisingly it was the case! Actually, the business has multiple websites that are for large markets, including Canada, the United States, United Kingdom, Canada, Italy, China, Japan, and others.

As the chief executive officer, Bezos did not permit his staff to make use of PowerPoint presentation. Additionally, he forced employees to write the form of reports that were six pages long to stimulate analytic thinking instead of basic bullet points. He was budget-conscious and established an organizational culture that did not offer incentives like free food or other luxuries.

Market Diversification

Bezos has not stopped diversifying the products of Amazon beginning with sale of video discs and CDs in 1998 before shifting later to toys, apparel and other electronic goods as a result cooperation with retailers and other companies. A majority of dot.coms fell in the beginning of 1990s. In a surprising way, Amazon experienced a boom in sales, which grew from just $510 000 in 1995, to an astounding $17 billion in 2010 plus more in 2011.

In the last quarter in 1999 Amazon became a multi-billion-dollar conglomerate that had 3.5 million label sets for auction. As time passed, it grew into the market of various goods that included video and music tapes. Jeff was able to start business operations in Europe. The total wealth of Jeff was estimated to be around $10.5 billion. Amazon businessman had reached around $10.5 billion. One of the most significant accomplishments of his career during 1999 was his being named

"Person of the Year" by Time Magazine and named the "King of Cyber Commerce."

However, the huge successes there were some downfalls. Amazon.com became involved in a variety of litigations brought by two major retail stores, American multinational Wal-Mart along with Barnes & Noble, a Fortune 500 corporation and online provider of digital and printed content. However, among the most significant cases was the loss of Amazon in 2000, which reached up to $1.4 billion. Following the year, the Amazon's management was forced to lay off 1,300 employees in a heartbreaking manner and enacted cost-cutting measures. However, the adoption of cost-effective policies helped Amazon and Amazon achieved its first profit net of $5.1 million during the final quarter of the year.

Incredibly, Jeff's schedule didn't stop his from putting aside time to read. Jeff, the Amazon owner is a book lover and purchases ten books each month, but is able to only read three. The books he favored included The novel by Frank Herbert Dune, Kazuo Ishiguro's The Remains of the Day, and Built to Last: Successful habits of Visionary Companies composed by James Collins and Jerry Porras.

Since its inception, Jeff and his wife are determined to turn Amazon.com into a online store where customers can purchase anything they want. Everything! Incredibly, their interest in creating a successful company has received the similar attention to their charity work.

Amazon Prime

Image retrieved from:

https://www.amazon.com/gp/feature.html?docId=1000991791

Additionally, in 2004 Amazon purchased the Chinese online bookstore, Joyo, in for 75 million. In 2004, with its new name, Amazon China, and stature that the bookstore was transformed into its seventh regional website. In February of the next month Amazon Prime, a multi-channel express shipping platform featuring more than one million items created. The concept behind the program was to create rapid delivery an everyday event instead of an intermittent satisfaction.

Through Amazon Prime, Bezos initiated three new innovations: the two-day delivery offer, that saw the majority of customers paying up to 6 business days. and a promise on the 2-day delivery, and inexhaustible. The first year was the time that Amazon.com was forced to sacrifice hundreds of millions in the form of shipping revenue. The program later produced positive results since a large number of customers enrolled in the

Program. In actual it is estimated that four of five initial members are still in the program until this time. Actually, Prime is still the most popular subscription in the world and millions of Prime members enjoy fast and unlimited delivery on over 30 million items.

Chapter 18: Stingy Philanthropist

In 2012, the question was posed for the people to ask "What will the Amazon baron do with all his riches?" It is likely that they are unaware of the charitable efforts of the patron. In the same time, Jeff Bezos gave away an enormous amount of fortune ($42 million, to be precise) along with part of his Texas property to build The Clock of the Long Now that is an underground mechanical 10,000-year clock, which is expected to run over 100 years. The company also donated $.5 million for the defense of homosexual marriage within Washington, DC.

What Charity Means to Jeff Bezos

Bezos is a man of insight and revels in lavish lifestyles, but Bezos has been criticized by critics for his incredibly tight-fisted. Actually the story in The Seattle Times in 2009 cried out that "lemonade stands will donate more to charity" in comparison to the Amazon owner. Amazon. The article noted that Bezos offered local associations for writers small amount, however these were vast difference from a business that has cash stocks that is worth $5 billion, and is earning $1,500 each minute of every day.

In spite of the fact that he was portrayed as a miser the people who were who were close to Bezos thought this was the best way to be sure that his funds go to the correct cause, the person or organization.

In 2003, family members established an organization called the Bezos Family Foundation which was subventioned through Amazon stock and administered by the parents of Jeff. It is impossible to find his name on any of the foundation's legal documents.

Evidently, a lot of people are unaware about the charity efforts of Amazon. Amazon creator.

One of the most significant contributions Jeff made to the healthcare sector was the $1 million that he contributed to the Fred

Hutchinson Cancer Research Center to support a program designed to expand the usage of particular immunotherapy strategies. The treatments targeted the prostate, breast and ovarian cancers.

The couple also made a pledge of $15 million to the university to establish the Bezos Center for Neural Circuit Dynamics. The center focused on an emerging field of research that was dubbed "Connectomics" and contained functions that included the study of neural connectivity, as well as mining data to improve understanding of human brains as well as the treatment of neurological diseases. According to reports, both spouses contributed another $20 million to research in the treatment of immunotherapy.

The family foundation has donated massive sums of money to both large and small educational grants. The foundation funds for the Bezos Scholars Program in Aspen Institute each school year. An interesting aspect with this is the fact the fact that the funds come from via shares Jeff donated to his parents, not his own fortune.

NO Political Motives

PayPal the founder, billionaire and PayPal who is also a politician activist Peter Thiel as well as the Canadian-American business mogul, Elon Musk, spend their money on political campaigns. Jeff Bezos is exactly their opposite. In the few occasions he stepped into politics, he made donations only in tiny amounts. His favorite politicians included Republican Congressman Jason Chaffetz, during his campaign for re-election in 2015. Jason won against Stephen Tyron, former executive of Overstock.com and a long-standing competitor to Amazon.com. Jeff has also supported the Political Action Committee of Amazon and Patty Murray's campaigns. Democratic Senator Patty Murray.

Bezos wasn't an active person. Maybe this is the reason why his opponents believe he lacks the direction of others or not interested in the world of politics or philanthropy. To counter this perception, Jeff tweeted and asked supporters on how he would distribute the funds of his own personal account. The exact message was, "I'm thinking I want much of my philanthropic activity to be helping people in the here and now--short term--at the intersection of urgent need and lasting impact." The donation was a big hit.

At some point Jeff Bezos, the founder of Amazon.com might have surpassed Microsoft's Bill Gates as the world's highest-income person (although the billionaires' list always changes). But it's likely that Jeff Bezos can ever become the most generous person anywhere in the world. In fact, if you insist on him being so, it will actually be untrue. Despite his claims to fame as a billionaire and raking in over $90 billion this man still has to discover what it is to be generous in the framework of humanity's understanding of Philanthropywhich is largely dependent.

Jeff is a member of the board for The Bezos Family Foundation which is focused on education for pupils in the school system. As of 2015 Jeff's parents gave around $68million (company stocks) towards the foundation in accordance with the latest taxes filed with the IRS. Jeff contributed $6 million as stock to the foundation.

New Brand of Philanthropy

It is certain that Jeff Bezos ignited a transformation or revolution within the world of retail. However, when charitable giving is concerned, he's just a tiny voice out in the middle of the night. Friends however claim that their friend is a nanny to Amazon and is likely to, in the end, switch towards altruism someday nearer the time!

There's no doubt about it that, like many other tech pioneers the man has a clear goal and that's not just to earn money but also to please the current generation of customers. This was evident in his famous line. fact. "I took the less safe route to follow my passion, and I am proud of making that choice."

However, Bezos savors the life of billionaire. In the past, he bought the most expensive house ever purchased at Washington, D.C. The buyer paid $23 million to purchase the Textile Museum that is one of George Washington University Museum. According to a few people who are close to Jeff his plans are to turn the museum into a one-family homes. Additionally, he owns a massive home located in Seattle with a price tag of $24 million. home located in Beverly Hills, California, as well as a huge ranch located in Texas. Jeff is believed to be among the most powerful landowners in the USA having more than 300,000 acres of terrain.

While Jeff Bezos is not like the other wealthy Americans The family has strengthened its sense of generosity over the last several years. In 2010, he stated to his long-time Television talk host and journalist Charles "Charlie" Rose, that, in the majority of cases these platforms are for profit and can improve the quality of life in comparison to the philanthropy models when they are harnessed to function effectively. The effects they have on society are both real and fundamental.

Problem in the face of Donald Trump Months before Donald Trump was elected as the candidate of Trump's Republican Party as its presidential candidate, Trump hurled a tirade at Bezos, the Amazon businessman for what he claimed was tax evasion, and the abuse of power and being a newspaper's owner, he could make personal attacks and impede legislators from investigating Amazon. Bezos denied that the accusations of Trump are unsubstantiated and do not make a good impression for a person trying to be the most powerful person on the planet. Bezos suggested Trump to act accordingly, even when Jeff Bezos said there was no reason for the former businessperson changed into a politician to look at and denounce his company.

He took part in a discussion with the President who was newly elected to the

United States of America and the world's leading technology companies in November 2016. At that time, Jeff Bezos uttered, "The Trump Government can become an innovation administration." However, that may not be the way Jeff is feeling today after completing an abrupt turn and pledging to fight legal efforts by the president to block travel to the USA in a number of nations including those that have significant Muslim populations.

In comparison with Bezos with respect to his wealth of Trump's net worth Trump dropped by $800 million over the past year. Property assets in the real estate sector that Trump was not orthodox floundered since retail prices suffered due to the growth of Amazon.com. Forbes revealed that Trump's worth was $3.1 billion, which was compared to $3.7 billion the prior year. The US Trump's position also dropped from #56 to #48 in Forbes' Forbes List.

Chapter 19: What Lies Ahead

If you're looking at Amazon.com It appears to have an extremely bright future, unless the business' fortunes fall as Jeff Bezos suddenly decides he is ready to cut his losses.

The efficiency of logistics is key to Amazon's continued profit. Amazon spends billions of dollars every quarter on shipping. Its costs are increasing as it strives to provide nearly every item in the consumer market, from toilet paper, paper towels and even paper and consumer electronics in 2 days or less customers. Amazon may be considering taking over duties of shipping for international couriers and delivery companies such as FedEx, UPS and DHL through leasing cargo planes, trucks and ships. A drone test could be a great idea in conjunction with this project.

Image retrieved from:

https://www.amazon.com/Amazon-Prime-Air/b?node=8037720011

In its focus on Alexa as well as AI (Artificial Intelligence), Amazon gained substantial advantage over rivals like Google and Apple on the world of virtual assistants. But, Google is breathing on its own neck with the creation of Google Assistant to be used with the Google Home automated device.

Amazon India

"India Experiment "India Experiment" was a huge accomplishment. Amazon continues to make investments into the country, and is planning to invest an additional $3 billion on expanding the services it offers. Just last year, Amazon.com unveiled its Prime Membership program in India as well as a plan to duplicate the success of the company in the US and in other regions. The company's management had previously stated that it would be expanding Prime Video to its subscription procedure, which will give customers who are in India access to video content featuring local actors and motion-picture creators.

Amazon.com has been subject to criticism by critics who harp on the company's overzealousness as well as its plans to control the entire world. The company was further strengthened by the purchase from The Washington Post. In the following months He also bought Whole Foods, the American supermarket chain, at $13.7 billion. In the meantime, it's been reported on the rumours that the next target of his is Nordstrom, a chain of luxury department stores called Nordstrom located in Seattle.

Are he also setting his sights at Apple or perhaps Cherry Hill and Orange County? It's only time to find out.

Since Bezos purchased Whole Foods, the public is clamoring at his latest innovative concept in the world of modern retailing.

Today, customers can place orders for groceries using Amazon Echo. Amazon Echo Personal Assistant (PA) or, as it is known, the "dash button." Orders generally take between a couple of hours and days, depending on quantity and availabilities. And then it was"Echo Fridge "Echo Fridge" came out from the blue, and resolved the issue.

The refrigerator is likely to have doors that face to the exterior. It can be as large or smaller as an air conditioning device. Two doors will open that open to this Echo Fridge from the kitchen on the other side. This is the main change in this crazy concept proposed by Argodesign which is a innovative and design-oriented company.

Amazon management has announced that its their online sales for Whole Foods reached $26.4 billion following the completion of its acquisition of the grocery chain with an increase of 22% year-on-year.

Like Steve Jobs' Apple, Amazon has become too popular and it is in need of reviving its massive profit, after its sales in the second quarter of this year grew by 25 percent to nearly $38 billion. Why? There is no way to stop people from purchasing every item they can on the internet.

An efficient and flawless layout will allow Amazon.com to stay ahead of growth and development while being prepared for the inevitable consequences from AI and mobile technology Amazon.com is focused to improving and improving.

Check out this fantastic strategy:

The future-oriented headquarters (HQ2) located in an urban hub or suburb which boasts of a powerful technical capacity as well as a minimum 1 million inhabitants. It will be in addition to the current headquarters of Amazon located in Seattle (Washington Washington). A minimum of 50 000 employees are expected to be employed and will receive an impressive salary as the management promises. The decision by

Amazon can create numerous employment opportunities at the location.

With no hesitation cities will be looking the eyes of Amazon for more than just a business partner due to its success CULTURE. That's the reason Amazon opens its to the.

Alpha of Retail

The 3rd of February last year Forbes published an article that reads, "Amazon Go is the Future." ..." Writer MY SAY wrote that "Amazon Go is currently being tested, however, it is Amazon's goal to be the ultimate Alpha of the retail industry. The giant is prepared to take your cash regardless of whether you're using them on the internet or in person."

I'll give you the exact description of:

"Amazon Go" is the newest style of retailer that does not have payment required. We have created the world's modern shopping system to ensure that you do not will have to stand for a long time in a line. Our Just Walk

Out Shopping experience just utilize Amazon Go, the Amazon Go app to enter the retailer, select the items you're looking for then leave! No lines, no checkout. (No, seriously.)"

What is the mechanism behind it?

This is a no-checkout shopping experience facilitated by technology that are used in self-driving vehicles, and assisted through computer vision sensors fusion and deep learning. This Just Walk Out Technology will determine automatically when products are taken out or placed back on the shelves, and will monitor the items using a virtual carts. When you've finished shopping, shoppers are able to leave the shop immediately.

In the next step, Amazon will charge the account of the customer's Amazon account, and then send an invoice to the customer. Jeff Bezos plans to spend $5 billion to building and design of Amazon's headquarters.

Consumer Engagement

The Amazon trademark is a source of consumer engagement with ease, comfort and convenience. In reality, many novelty products in the retail market fall short due to the lack of commitment and acceptance.

Let's look at the instance for example of Macy's Department Stores which is an iconic shopping spot located in New York City. Macy's Department Stores has lost its prestigious position in part due to its product range that some people have noted.

The situation at Amazon is totally new as it currently is home to 54 million customers and 244 million customers who are active. Are you aware that 70 percent of American households have purchased products at least once through the most popular online retailer around the globe? People will never be hesitant to download apps once they figure out that it's from Amazon.com. What is more important than this? I want to know!

United Kingdom Amazon

Amazon has fulfilled its promises towards its customers in the United Kingdom when it opened an development center in the center of University City of Cambridge (Eastern England). The structure is made up of three levels of a luxurious office, which houses around 400 workers who conduct studies on products from Amazon Assistant (Alexa) to the new area that includes Prime Air deliveries, making the use of aircrafts that are not piloted.

The new development comes in an addition to the initial Castle Park structure also erected in Cambridge for a team comprising scientists, engineers analysts and researchers (Amazon Research Cambridge) that will focus to develop "real innovation", as stated in the Amazon's United Kingdom head, Doug Gurr.

LESSONS FROM THE LIFE OF JEFF BEZOS

Jeff Bezos, the Amazon.com's chief executive, is a mastermind in a class that is his own. He has demonstrated to the world that with a keen eye and effort, anything is only the start.

He has had jaw-dropping huge successes thanks to his ingenuity and bold concepts. In a few pints his name was on Forbes listing of the most successful billionaires around the globe and beat Microsoft's CEO, Bill Gates many times. In November 4, 2017, for example, Jeff Bezos total net worth was $94.6 billion in accordance with information obtained from Bloomberg Billionaires Index. Bloomberg Billionaires Index. The feat was not achieved from a series of random events. This is the result of unending determination and energetic dedication. This is where I look at some insights from the experiences of this successful entrepreneur. Also, I will look at the way he's handled his family's needs and the immense responsibilities that the man is responsible for.

1. Your customers should be the focus rather than rivals

Most businesses, particularly the ones in the technology or online commerce, concentrate on their competition instead of their clients.

This is a major omission that we've learnt from Bezos approach to organizational procedures. In one of his writings, "Many companies describe themselves as customer-focused, but few walk the walk. Most big technology companies are competitor focused. They see what others are doing, and then work to fast follow." It is important to focus on the customers you serve. This provides you with an edge and puts you an extra step in front of your rivals. Keeping track of your competition creates a copycat who is constantly looking to improve upon the ideas of other companies. Consider the Amazon Web Services (AWS) as an example. Bezos created it to address the issues of expensive internal application hosting, as well as open-source solutions that can meet the needs of growing and/or already huge firms. The idea would've been impossible if he had been focused on the competition. He was attentive and read the feedback from his clients!

2. To be a leader, you must take similar risk

There is nothing worthwhile or cherished that can be made without risk. Most of the most significant achievements and innovations were based on risk. Bezos is no different. In various times the Amazon founder took breath-taking risk to establish himself as the leader on the marketplace. To become the first worldwide online retail location of choice Bezos attempted to take risks that failed only to try repeatedly. The first time he tried it was with Amazon Auctions, then zShops followed by Amazon Marketplace which accounts for nearly half of all the items currently sold through amazon.com. Bezos said in his initial annual letter (1997), "Given a 10 percent chance of a 100-times payout, you should take that bet every time. Failure and invention are inseparable twins. To invent you have to experiment, and if you know in advance that it's going to work, it's not an experiment."

3. Give your employees the power to be empowered, and give them the wheel

One of the most important factors in the performance for a business is the dedication and trust of employees. It is crucial to break relations with employees. It is also important to give them strategically an ownership stake in the company. They should be connected to the success of the company. Bezos has shattered the yoke of options to purchase stock when hiring. He in 1997 stated in his initial Amazon annual letter "We will remain focused on the hiring and retention of highly skilled and flexible employees and we will continue to tie their pay in stock options instead of cash. Our success will largely depend on the ability of our company to recruit and maintain a committed workforce, all of whom will think as an owner and, therefore, be an owner."With this approach employees aren't seeing the work they are required to perform every day, but instead their life in which they are required to live a fulfilled.

4. Create a company style and follow it.

Bezos has proven that in order in order for a business to succeed the company must create an individual culture that takes into account its uniqueness. The culture does not require the highest or the most appropriate collection of principles to follow but it should be one that can bring the best results, effectiveness and efficiency. Bezos stated in his 2015 letter. "We never claim that our approach is the right one -- just that it's ours. Over the last two decades, we've collected a large group of like-minded people. Folks who find our approach energizing and meaningful." The other key aspect in creating your unique corporate approach is input from other like-minded individuals. It is very valuable if it's done.

5. Eliminate the process of decision-making for avoiding unnecessary bureaucracy.

For quick progress and to speed up the pace of action for important tasks The decision-making process for a business have to be fluid and flexible. Important and crucial choices could be left up to directors as well as high-ranking executives. However, other employees should be able to take quick judgements and decision-making. This is one of the secrets to Amazon's amazing achievements. Bezos has even written in his paper of 2015 on the kinds of decisions that should be taken and the people who must decide in which contexts and in what timeframes. The first of these he refers to as "Type 1" decisions," which should be made by the senior executive. These are crucial to the survival of the business. He affectionately refers to the Type 2 decision an option for employees in general and could be changed. "Type 2 decisions can and should be made quickly by high judgment individuals or small groups. As organizations get larger, there seems to be a tendency to use the heavy-weight Type 1 decision-making process on most decisions, including many Type 2 decisions."

6. Do not delay. Get it done today!

The way Jeff Bezos began his Amazon journey will always be an inspiring and motivating. The founder was trapped between whether to quit an employment that was lucrative and launch the business of his own or to stay. He once stated, "If you decide that you're going to do only the things you know are going to work, you're going to leave a lot of opportunity on the table." That means that it's best to act when we are able to come up with a great idea. The idea doesn't need to be developed fully. It's more likely to lead to inaction, or putting off work. Bezos has often encouraged his employees to be prone to mistakes when they make efforts because it is more effective than being inactive. The founder of Amazon said that "We are willing to go down a bunch of dark passageways, and occasionally we find something that really works." The six main goals of Amazon include "bias for action" as an element. It demonstrates that action is profitable and can be at the heart of Bezos his success story.

7. Reduce regret to the minimum is

In making important decisions about your future when making life-changing decisions, it's more crucial to conduct a thorough review of your current circumstances as well as the possibilities. Any plan that doesn't consider reality is certain to be a failure. As Bezos was considering quitting the company, he was weighing the choices. It was his intention to not wish to make an unwise decision. The idea was that he would never choose a different one that would cause him to look back on 20 and be remorseful about the chance that he was unable to take. He stated, "The framework I found that made my decision simple was what I call"what only a nerd could call a'regret minimalization framework'. Therefore, I decided to bring myself into the future to be around 80 and think, "Okay and now I'm reflecting on my entire life. I'd like to reduce my regrets that I've got."

8. The slow and steady win the race

If you started a business in the past, how long will be required to begin making real-time profits? I bet you'll think two years, or an entire year or lesser. The time it took Amazon six years, and in that time they only made 5 million dollars in profits on revenue that was more than $1 billion. But they didn't stop. The slow pace was an integral part of Bezos strategy from the beginning. The company was designed to be cost-effective and reinvesting the profits in the highest amount. The stakeholders and investors were initially difficult, but soon started having fun with the surge that ensued. This is an invaluable knowledge. It is possible that your start-up has several challenges during its early times. This doesn't mean that you must end your business. The only thing you have to do is be able to conduct an study of your circumstances then create a viable program, adhere to your plan, and develop gradually but surely.

9. Get a lower price!

Numerous businesses make the error of increasing their costs without having a corresponding increase to the standard of their product or services. Although the service or products are improved however, it's still ideal to keep prices as low as you can to increase the customer base, and earn greater profits over the end. It has been proven the ideal approach. In fact, charging more could cause an increase in revenue, but in the end, it proves to be inefficient. This could result in losing loyal customers that look for alternative options that are more efficient and less expensive. This is always going to happen. That's the way of doing business of Amazon. In addition to constantly improving their skills, they seek ways to make prices at a minimum. Bezos has said that, "There are two kinds of firms: the ones that are trying to make more money as well as those who are working to lower their prices. There is a second." The fact that it is a certainty is what has pushed him into in the world of e-commerce, expanding his customers.

10. The pace of innovation never stops.

"What is dangerous is not to evolve" -- Jeff Bezos, CEO & president of Amazon.com

At some point, we all have a fear of changes. So long as we are familiar, we avoid any effort to shake things up. The real success lies in experimenting with new ideas. There is no way to test new ideas when you don't test the new concepts. This is the guideline of Amazon worthy of imitation. Amazon was founded with selling of books. In the present, they have grown to sell almost anything in and out of other products including delivering groceries right to the door (if you reside somewhere in Washington). It is interesting to note that they continue to change and that's what has made them successful in their and success.

11. Do not reside in Utopia

It's great to be creative. The ability to think creatively isn't from the norm. In fact, some of the most innovative ideas originate from deep thought. But one has to remain sensible. Being unrealistic has been an issue for a lot of entrepreneurs. It's led to the demise of numerous companies. It's a learning experience Bezos can teach with his companies. It's good to dream large-scale goals for your business however, you need to consider what reality is. Most entrepreneurs are ambitious and fail to consider, or invest the same amount of effort and ultimately fail. Bezos brilliantly demonstrates this in his statement, "It's very important for entrepreneurs to remain realistic. If you think the first time you're writing your business plan, that there's a 70% possibility that your entire venture could fail, then it alleviates some of the stress of doubt." The mindset of this type helps cushion the repercussions of business failures or inefficiency. The entrepreneur is able to take the time to prepare for every possibility.

12. Are you able to see ahead?

For you to achieve great success in your business, you need think beyond what is happening now. Being able to evaluate your present situation to the future and take concrete business decisions will ultimately make you stand in the end. The issue that entrepreneurs who are shortsighted face is that they become out of time as the future that they haven't prepared for arrives. The result is usually closing or becoming irrelevant. This is because Amazon chief, Jeff Bezos also recognizes the fact that this is a constant. The Amazon boss said "a large majority of people think that it is best to live in only the present moment. However, I'm not among the majority of them." It's an essential principle which has kept him in the spotlight. In the case of Amazon one might have imagined that in the 80's the 80's, that millions of people could easily at the ease of their home, make an place an order and receive the item almost instantly? Bezos looked at the trends and forecasted what was to come. It was true, according to reports!

Chapter 20: Who Is Jeff Bezos?

The Washington Post is owned by the businessman and e-commerce pioneer Jeff Bezos, who also established the space exploration business Blue Origin and is the chief executive officer and founder of Amazon. He's among the top millionaires in the world thanks to his successful business ventures. Bezos was born in 1964, in New Mexico, developed an early fascination with computers. He went to Princeton University to study electrical engineering as well as computer science.

Following his graduation after graduation, he began his career on Wall Street until joining D.E. Shaw in the year 1990, as the company's the youngest vice president. Bezos quit his job in the year following to create Amazon.com which is an online bookstore which was one of the biggest successes on the Internet. 2013 saw Bezos buy The Washington Post, and 2017 saw Amazon buy Whole Foods. Bezos is set to retire as the CEO of Amazon at the end of the second quarter of 2021. the company announced in February 2021.

EARLY LIFE

The birthplace of Jeff Bezos was Albuquerque, New Mexico, in January 12 1964 Jeff Bezos was born to Jackie Bezos, who was only a teenager. It was just over a year ago that Bezos's parents got married. In the year Bezos his son Jeff was just four year old, the mom quickly got married to Mike Bezos, a Cuban immigrants.

Bezos was awestruck by the ways in which things operated when he was a kid turning the garage of his parents to a laboratory and building electronic devices throughout the house.

Jeff confidently stated that the human race's future lies beyond Earth during his classes on space and space exploration in the school. Jeff launched his first company in that year and also. "The "Dream Institute" was the title of the summer school that was offered to students who were in fifth, fourth and sixth grade.

The pupils who participated to the school were taught a lot of reading material, and its goal was educational.

EARLY CAREER

Bezos received offers of employment from Bell Labs, Intel, as well as Andersen Consulting, among others when he earned his university degree in 1986. The first company he worked for was the financial telecom startup Fitel and he was charged with establishing an international trading network. Following that, Bezos was elevated to the position of director of customer service and director of development. Bezos changed jobs and entered the world of banking in 1988 after being appointed a product manager for Bankers Trust. Then, he joined D. E. Shaw & Co. in the year 1990 which was a new hedge fund which was heavily was a mathematical modeler. The company he worked for until the year 1994. He was 3 years old when Bezos was appointed D. E. Shaw's 4th vice president of seniority.

FAMILY & PERSONAL LIFE

The family moved into Houston, Texas, when Mike completed his studies at the University of New Mexico so that he could begin his career as an engineer with Exxon. Jeff went to River Oaks Elementary School in Houston between the fourth and the sixth grade. Lawrence Preston Gise, a regional director with the U.S. Atomic Energy Commission (AEC) located in Albuquerque was Jeff's maternal grandfather. Jeff had several summers when a kid on the ranch of his family located in Cotulla, Texas, where Lawrence was laid off at an early age.

Then, Jeff would buy this ranch, and expand its size to 25117 acres (10,117 ha) to 300,000 acres (121,406 ha). To prevent the younger children from entering the room Jeff had a time modified an electrical alarm. Additionally, he displayed his technological skills and curiosity. Jeff went to Miami

Palmetto High School following his family moved from Miami, Florida. Miami, Florida.

Bezos married Mackenzie Tuttle, whom he was acquainted with at D.E. Shaw, in 1993. They announced their separation in January of 2019 on the same day that The National Enquirer published a report in which it claimed that Bezos was involved in an affair with another woman. The publication's discovery of Bezos's private messages led him to launch an investigation.

In February, he wrote an essay that was long on the Internet where he alleged officials from American Media Inc. (AMI) who publishes the Enquirer of "extortion and bribery" for the alleged threat to release naked photos of Bezos in the event that he didn't stop his probe and threatening other demands. In the investigation led by Bezos these texts were published by his brother's lover.

WIFE AND KIDS

In their workplaces The hedge funds D.E. Shaw The hedge fund's D.E. Shaw Bezos as well as his ex-wife Scott later referred to as Tuttle were married in 1993. It was also in the same year he launched Amazon. In 2019, the couple got divorced. They have now divided the control of the four children 50/50.

Anchor of the news Lauren Sanchez is the person Bezos has been seen with.

Jeff Bezos and MacKenzie Scott have three sons and a daughter. The daughter was taken in by China and adopted by their couple. While Bezos is known to prefer keeping his children's names private, there are rumors that his 21-year-old son's birth name has been changed to Preston Bezos.

The oldest son of the couple, Preston, is now in a separate residence from the couple. Similar to Bezos and his wife, Preston is was enrolled in Princeton University. Preston is known to like dining out Mexican cuisine and travel.

Make your own decisions, rather than the capabilities you possess. When he was interviewed, Bezos was asked what tips he'd give his children. To which the answer was, "You've got to figure out what you love, and it's going to bring you enormous joy."

The future heirs of Bezos the fortune of billionaire Bezos will consist of all four siblings of the founder's children. MacKenzie got 25 percent of the shares at Amazon due to his divorce from his former wife and Jeff retained 75percent of his shares. Additionally, Bezos gifted MacKenzie 4 percent in his own business.

WEALTH

His wealth is more than those of Microsoft Corp. co-founder Bill Gates and famous investment guru Warren Buffett as of late 2021, Jeff Bezos is now second richest on Earth. Bezos has a net worth of in the moment of date is believed to be $197 billion, according to the Bloomberg Billionaires Index.

Cloud computing is a booming system Amazon Web Services, the market leader in online shopping will only continue to rise up the ranks, which is good news for the founder of Amazon Web Services as digitization transforms the way we behave and the revolution in cloud computing changes the way businesses operate.

An online platform which could represent up to 9percent of transactions in retail stores across the United States and a staggering 51.2 percentage of total online shopping in 2020 was run by Amazon's founder the company's former CEO as well as its the current chair of its executive committee.

Bezos the well-meaning manager tried to convince him not to quit his job in D. E. Shaw & Co. at the time Bezos came up with the concept of his online-based business. Bezos was raised by his mother as a child and then by his Cuban immigrant stepfather, has always was looking to develop something different.

Chapter 21: Business Career

FOUNDER AND CEO OF AMAZON.COM

Amazon.com was founded around 1994, with the help of Jeff Bezos. The most customer-oriented business in the world Earth is the goal of Amazon. Amazon produces and makes the bestsellers Kindle, Fire, and Echo products as well as its Alexa voice recognition system. Amazon offers a wide selection of movies and TV shows via Prime Video, offers low costs and fast delivery of thousands of products, and supplies governments and companies across 190 countries with the top cloud computing platform by using Amazon Web Services. Bezos created the aerospace firm Blue Origin, aiming to ensure that space travel is safe as well as less costly. He also controls Washington Post. Washington Post.

AMAZON INSTANT VIDEO & AMAZON STUDIOS

The leader in the field of cloud computing and e-commerce, Amazon.com Inc. (AMZN) is one of the largest businesses worldwide. In July

2022, the market capitalization stood at $1.25 trillion. It was 1994 when Jeff Bezos founded Amazon as a bookstore online. The company quickly was able to become a key market in the field of e-commerce. It offers diverse products like electronics as well as furniture, clothing food items, toys and much more.

In the following years, Amazon expanded its business to cloud computing streaming video and music groceries, as well as the creation of content. Amazon also manufactures and sells consumer items such as the Kindle as well as the Echo. Amazon has reported net revenues of $469.8 billion and net earnings of $33.4 billion for financial year 2021. These figures increased by 22 percent and 56% over the prior year's results.

KINDLE E-READER

Amazon has created and is selling the line of digital reader devices under the brand name Kindle. Wireless connectivity allows customers of Kindle devices to shop, browse or download online newspapers, books magazines and other items from the Kindle Store. The hardware platform was created by Amazon Lab126, a subsidiary of Amazon Lab126 and made accessible to users as a single device 2007. It includes a range of devices which includes Kindle software, which is compatible with the majority of computer systems as well as e-readers that have E Ink electronic paper displays. Its Kindle Store works with all kindle devices, and provides over six million ebooks available in the U.S. at the time of writing in March 2018.

AMAZON DRONES

A drone-delivery service, known as Amazon Prime Air, or just Prime Air, is currently being created by the retailer. It uses delivery drones to transport packages to customers independently. The FAA selected the company to take part in a form of delivery certification program that includes drones by 2020. The program was jointly alongside Zipline, Wingcopter, and seven other firms. At

Lockeford, California, activities are expected to commence by 2022. the end of June 2022.

Amazon subsidiary Lab126 has developed the hardware platform.

The company's ongoing U.S. testing, a one-time drone delivery cost minimum $484 in 2022. Amazon expected the cost to fall to $63 by 2025 however, it would be approximately 20 times the price of the typical cost for ground delivery. The customers who participated in the trial in the past had to install a physical mark on their property to mark the location of drop-off as well as select someone who would be able to observe the drone's flight path. The drones used had an area of five kilometres, and cost 146,000 each to build.

FIRE PHONE

Amazon launched its Amazon Fire Phone for smartphones during July of 2014. The device comes with an 4.70-inch display with 315 pixel density, and has a resolution of

720x1280 (PPI). The Quad-core Qualcomm Snapdragon 800 processor powers the Amazon Fire Phone with a 2.2GHz clock speed. There is 2 G.B. of RAM available. There is a lot of RAM available. Fire Phone is powered by 2400mAh battery that is non-removable and is running Fire OS 3.5.0.

In terms of photography In terms of camera, in terms of photography, the Amazon Fire Phone boasts a 13-megapixel camera in the rear. It also has a front-facing 2.1-megapixel camera that can snap selfies.

32 G.B. of internal storage is available in Amazon Fire Phone. 32GB of storage capacity is available. Amazon Fire Phone, which operates on Fire OS 3.5.0. The Nano-SIM SIM card is employed in conjunction with this Amazon Fire Phone, a one SIM (GSM) phone. The mass of it is 160.00 grams. Amazon Fire Phone is 160.00 grams and the dimensions are 139.20 inches x 66.50 8.90mm. 8.90mm.

Bluetooth 3.0, NFC, 3G Wi-Fi 802.11 a/b/g/n/ac and 4G are some of the Fire

Phone's available connectivity. The phone has a proximity sensor, gyroscope, barometer, accelerometer, ambient light sensor, compass/magnetometer, and other sensors.

Amazon Fire Phone prices in India start at around Rs. 10,150 as of November 4, 2022.

WHOLE FOODS

Jeff Bezos purchased Whole Foods Market for $13.7 billion in August 2017.

Amazon tried its hand at to enter the market of grocery by launching a product known as HomeGrocer. But, it didn't grow. The opportunity to acquire Whole Foods presented itself, it was difficult to know whether the platform's e-commerce was planning to go back to its business.

However, given that $13.7 billion isn't an insignificant amount, Amazon must have understood the importance of this sector. be.

AMAZON WEB SERVICES (AWS)

In the year 2006, Amazon Web Services (AWS) began to provide companies with I.T. infrastructure solutions in the form of web services. They is now known by the name of cloud computing. The main benefit for cloud computing the possibility to swap out initial capital expenditures for infrastructure at cost variable costs that increase with the business. Thanks to the Cloud cloud computing, businesses do not have to plan the purchase of servers or other I.T. equipment for weeks or months ahead of time. Instead, they're able to swiftly start up several hundred thousand servers, and provide results much faster.

A wider range of features and services are offered by each of the services at AWS more than any other cloud service, that range from the most basic technologies such as storage, computation databases and storage to advanced technologies such as artificial intelligence the machine-learning process, data lakes as well as the Internet of Things. It makes the transition of your existing applications to cloud computing quicker, easier as well as more cost-effective. AWS permits you to build nearly everything you could think of.

With these kinds of service, AWS also offers the most modern capabilities. For those who want to select the best solution for your needs and to get the most value and efficiency, AWS, for instance has the greatest range of databases that are specifically created to work with various types of applications.

The biggest and most active community is AWS with millions of users and tens of thousands associates across the globe. AWS customers use AWS come from all sectors and with every kind of size which includes large corporations, startups and even governments. Tens of thousands of software developers who are independent (ISVs) as well as thousands of system integrators who have an emphasis on AWS services form AWS Partner Network (APN). AWS Partner Network (APN).

Many thousands of businesses use Amazon Web Services' highly reliable, flexible cloud infrastructure that is accessible across 190 nations. Clients from all sectors profit by the following benefits thanks to data centre websites across The U.S., Europe, Brazil, Singapore, Japan as well as Australia.

BLUE ORIGIN

The development of a commercial space station known as Orbital Reef will be facilitated through a collaboration between Amazon as well as its cloud computing division, Amazon Web Services, as well as a separate company established by Jeff Bezos.

Together with the the Colorado-based Sierra Space, Jeff Bezos' Blue Origin space company is among the initiators of the Orbital Reef initiative. This collaboration comprises Arizona State University, Redwire Space, Boeing, and Genesis Engineering.

Orbital Reef was awarded an one-time $130 million NASA grant in December of last year to further develop the idea of an orbital satellite that could help fill the space left after space station International Space Station is shut down by 2030-2031. To explore their concepts for satellites, two other teams, which were led by Nanoracks and Northrop Grumman, also received money from NASA.

The WASHINGTON Post The Washington Post is a daily newspaper that is published every morning within Washington, D.C. The newspaper is the most read across Washington, D.C. U.S. capital, and frequently ranks among the top newspapers around the globe.

By signing up for this subscription, you can access digitally the entire content accessible on The Washington Post Company website as well as The Washington Post Company mobile applications is available without restriction. It also offers each day Kindle issue from The Washington Post Digital Access and can be read using any Kindle electronic reader or smartphone using the Kindle application.

Paid Kindle subscriptions are risk-free and offer a trial time before the subscription starts. Find out more about subscriptions.

If you do decide to end your subscription the subscription, your Kindle subscription to the publication will renew. If you subscribe to a subscription that lasts 6 months or more and in any other case when it's required by laws, Amazon will send you an email notification of renewal prior to the expiration date the subscription. The email will also contain the price of renewal. If you pay the lowest rate that is available to customers on Amazon.com during the renewal Amazon will renew behalf of you. If the price of renewal changes the customer will be notified via email notice. When the renewal order is completed, you may change the details of your credit card or subscription info.

BEZOS EXPEDITIONS

Bezos Expeditions operates as a family-owned office (SFO) that oversees investment in private equity and venture capital for Bezos.

The company's 159 employees comprise finance analysts, risk analysts and market analysts, the assets that it manages are believed to amount to U.S. $107.8 billion (AUM).

Since its inception, Bezos Expeditions has invested more than 118 times in both charitable and non-profit ventures. The most prominent ones include:

Airbnb

In 2011 this round included unidentified investments from Bezos his family office. Its shares nearly doubling by the second day of its IPO on Dec. 2020 was among most significant ever. There's no confirmation of whether Bezos retains his shares in the company.

Twitter

Prior to when Twitter was a publicly traded company in 2008, Bezos Expeditions invested $15 million into the giant social media company.

Plenty:

The family's business took part in the vertical farming company's Series B funding, which brought in more than $200 million to fund the venture.

Unity Biotechnology:

In 2016, the company which aims to reduce the negative effects of age-related complications was awarded Series B funds from Bezos his family office.

Uber:

Prior to Uber's lucrative IPO in the year 2019 The family office also provided US$37 million for its Series B financing round in 2011. Uber is predicted to reach a valuation of US$93 billion by 2021.

Museum of History and Industry:

Bezos donated to the museum in Seattle an astounding US$10 million in 2011 for the creation of an "Center for Innovation" in the new building.

Princeton University:

In order to establish the Center for Neural Circuit Dynamic dedicated to the study of neurological diseases Alums donated $15 million to the college.

Bezos Expeditions has recently expressed the desire to expand into the South-East Asian market for e-commerce. In the month of October, 2021 it was part of a $87 million Series B financing round for Ula Wholesale eCommerce, a platform offering solutions to small-scale businesses grow the amount of customers they have. The other investors were Northstar Group, A.C. Ventures, Proses Ventures, B-Capital, Tencent, and Citius.

www.ingramcontent.com/pod-product-compliance
Lightning Source LLC
Chambersburg PA
CBHW071443080526
44587CB00014B/1969